I0030898

DESIGN ASPECTS OF POWER TRANSFORMERS AND REACTORS

Jim Fyvie

Published 2016 by Abramis academic publishing

www.abramis.co.uk

ISBN 978 1 84549 683 8

© Jim Fyvie 2016

All rights reserved

This book is copyright. Subject to statutory exception and to
provisions of relevant collective licensing agreements, no part of
this publication may be reproduced, stored in a retrieval system, or
transmitted in any form or by any means, without the prior written
permission of the author.

Printed and bound in the United Kingdom

This book is sold subject to the conditions that it shall not, by
way of trade or otherwise, be lent, re-sold, hired out, or otherwise
circulated without the publisher's prior consent in any form of
binding or cover other than that which it is published and without
a similar condition including this condition being imposed on the
subsequent purchaser.

Abramis is an imprint of arima publishing.

arima publishing
ASK House, Northgate Avenue
Bury St Edmunds, Suffolk IP32 6BB
t: (+44) 01284 700321

www.arimapublishing.com

CONTENTS

Introduction 5

PART 1: TRANSFORMERS
Design Reviews 10
Various Types 12
Specifications 14
Core Magnetising Flux 16
Flux Density 23
Magnetic Circuits 26
Windings 32
Winding Leakage Flux 43
Reactance 47
Tapchangers 65
Thermal Considerations 71
Magnetic consideration 82
Electromagnetic Forces 85
Electric Fields 87
Insulation Strength 94
Insulation Systems 105
Processing 138
Predicting Performance 140
Testing 142

PART 2: REACTORS 155
Series Reactors 161
Shunt Reactors 175

References 211

Acknowledgements 215

Introduction

The aim of this book is to provide a guide for students who wish to understand more about the design of power transformers and Reactors. It is assumed that the student has some knowledge about electrical and magnetic theory; although some relationships are revised. The basic design procedures are discussed, as is the use of modelling using equivalent circuits. There are also sample EXCEL program outputs which cover reactors. The book is organised in two parts. Much of Part 1 will cover detailed information that is relevant to both parts, Part 2 details information for reactors.

PART 1
Transformers are used to transfer energy between two or more systems that operate at different voltages. They are very efficient, close to 100%, have no moving parts and last over twenty years.

There are several types of transformers and they may be designed in accordance with many different specifications and to operate within many different boundary conditions. The operating boundaries such as maximum and minimum system voltages, system fault levels, impedance characteristics, tapping range and load power factor are all interrelated aspects of the design.

Calculations to predict the performance are carried out by using mathematical models; these represent the transformer characteristics from several aspects. The magnetic, thermal and electric field models are used to estimate the mechanical, thermal and electric characteristics within the transformer.

The most frequently used model is to analyze the transformer as a single phase equivalent circuit using a two port network arrangement, which suits the testing as the results can be verified by open circuit and short circuit tests. It is normally not possible to test with full MVA at the manufacturer's works.

A major requirement is to model the magnetic leakage field; this determines the reactance, the short circuit performance, the stray losses both inside and outside the windings, the tank and clamp temperatures, the hot spot winding temperatures and the style of magnetic shields required. Accurate assessment of this leakage field is essential in order to predict the correct performance of the transformer and finite element methods are commonly used to carry out this assessment.

The next most important parameter, which dictates the clearances, is the electric field and the insulation stresses throughout the structure. The distribution of these stresses will vary in accordance with the wave-shape of the voltages concerned. Power frequency voltages and long duration switching transients, will distribute evenly in accordance with the winding turns. Impulse type voltages, due to the nature of the fast rise time, will distribute initially in accordance with the distributed capacitance and finally, with the slower tail, will follow the inductive distribution. During the transition there may be oscillations depending on the capacitance, inductance and resistance values.

Last but not least is the thermal model of the transformer, most power transformers have liquid insulation which also provide cooling, removing heat from the core and windings and transferring it to external coolers. A hydraulic model is required to analyze all the internal temperatures at full load.

Transformers are expected to operate beyond their normal nameplate rating for short periods of time. The philosophy being that the cellulose insulation life is a function of temperature and if the unit is operated at low temperatures for a given period, then it should be able to operate for a period at some elevated temperature so that the average expected life is maintained: this elevated temperature has a limit.

Transformers are also expected to operate at higher voltages for a period of time, this will involve some over-flux factor and the cores should be designed with this in mind.

The limits to these over load and over-voltages are set by the allowable hotspots and gassing tendencies of the materials used. Care must be taken to prevent any situation where free gas bubbles are generated as they can cause insulation failure.

Over-voltage protection is required for power transformers to ensure that any external over-voltages such as lightning strikes are prevented from exceeding the test levels chosen for the equipment.

The performance of a transformer is verified by tests, and these will be listed under a separate section.

PART 1

TRANSFORMERS

Design Reviews

For large transformers it is necessary to discuss the terms and conditions of the contract, these usually involve negotiations on price and delivery and any penalties regarding tolerances on performance.

As the units become more complex and the specialists in the utilities or manufacturers reduce in numbers due to downsizing, or reorganisation etc., the need to understand the technical issues becomes more important. It is therefore prudent to conduct technical design reviews at the beginning of the contract, or even at the tender stage. This reduces the risk of any misunderstanding during the manufacturing phase.

Often an independent consultant is contracted to assist. It is very important for the manufacturers to know what limits their product range. The closer the specifications are to the manufacturer's ability, the more effort is required at the tender stage. It may not be apparent from the outset, and it would seem that it is obvious, but before any major commitment is made, it is essential to identify any shipping limits or essential dimensional parameters, to ensure that a suitable test plant is available, and the manufacturer can build the unit with existing equipment. Any special needs must be identified prior to the contract agreement.

The design review may follow internal procedures as laid down by the manufacturer or the purchaser. If there are no such procedures available, I would recommend "GUIDELINES FOR

CONDUCTING DESIGN REVIEWS FOR TRANSFORMERS 100 MVA AND 132 KV AND ABOVE", which is an excellent document produced as Brochure 204, by CIGRE.

Various Types

Step-up transformer – These are located at the generating stations, and are often referred to as Generator Transformers, because they are often directly connected to the generator windings without circuit breakers. The generator voltages are usually limited to around 20 kV, due to the nature of the insulation and the close proximity of the iron stator frames surrounding the windings. It is not efficient to transport large power levels at these low voltages. Considering a power output of 800 MW at 20 kV, the three-phase current would be 23094 amperes. If however the voltage is increased to 400 kV, the current is reduced to 1155 amperes, which can be carried by overhead cables. Step-up transformers are therefore very special, as one winding has a very high current, and the other winding has a very high voltage. It is prudent to connect a high current winding in delta and a high voltage winding in star, as will become apparent later. At these high voltages the power can be transmitted over large distances, this allows power stations to be built near the fuel source, either hydro, which is located in the mountains, or coal, which is near a coal field or nuclear, which is remote from the cities.

Super Grid Transmission transformers – These are located on the periphery of the main load centers and provide an intermediate step down stage to take the voltage into a sub-transmission level, normally at around 132 kV. These transformers are designed to operate at nominal currents, where overhead lines can be used. These are rated at 400MVA or 240 MVA, but they have high voltages on both sides. Since both of these windings are usually connected in star, it is convenient to design them as autotransformers.

Large distribution transformers – Some utilities consider 132 kV as being the upper level of distribution and some consider these as substation transformers. The function however is to reduce the voltage from 132 kV, mainly from an overhead line to a 33kV or 11 kV cable box where the cables can go underground to feed the 33 kV local network transformers. They can be rated between 100MVA and 60 MVA, some of the units can be designed with two LV windings each carrying half the MVA.

The smaller transformers 33 kV and below are distribution transformers and are usually bought as bulk items. They will not be addressed in this book.

Specifications

There are many National and International Standards that are used to guide the buyer in selecting the parameters for a specification. The International Electrotechnical Commission (IEC) has produced a series of specifications that may be used to assist both parties to agree on such a contract and covers many different types of transformer. There is a section in Annex A of IEC 60076-1, which gives an excellent guide to the information required at the tender stage.

Transformer specifications and standards take many forms and the most common of these are listed.

International Standards

IEC 60076 series. This series is continuously updated and new specifications are periodically produced. The main committee dealing with transformers is TC 14, which consists of a number of representatives from some 32 countries through their National Committees. Up-to date details of the activities can be found on the IEC web site under TC14.

National Standards.

These are standards that are controlled and written by National committees, the relevant committee for the USA is the IEEE Transformer Committee. These specifications are also updated regularly and the latest lists can be found on the IEEE website under Transformers Committee.

ANSI C57.12.00 – General requirements
- New section for loss tolerance
- New requirements for Dielectric tests
- Definition of thermal duplicate

ANSI C57, 12-90 – Test Code
- New section on load loss, No-load Loss
- Wide band PD/RIV measurements 1 hour test
- Sound Power level
- Loss test guides, Impulse test guides

EN – European Normalization Standards are standards relating to the European community and are adopted from the IEC series; they are usually bi-lingual.

Bsi – Standards are also adopted from the IEC specifications but only contain the sections written in English.

DIN – German National Standards

AS – Australian Standards are unique to Australia.

There are other Nations that produce their own standards, such as India and Japan.

There are also Utility Standards, which may be site specific and Manufacturers Standards which dictate the in-house rules for individual manufacturers.

Core Magnetising Flux

A magnetic field (Φ) is produced around a conductor when an electric current (I) flows through this conductor. The relationship between the strength of this field and the magnitude of the current is the reluctance (S) of the path of the magnetic field. In its simplest form, as shown in Figure-1, for one conductor or one turn; $\Phi = I/S$.

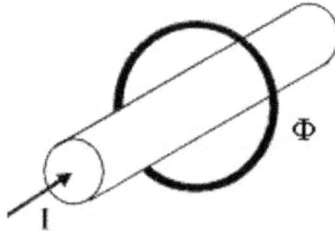

Figure -1

The reluctance is a function of the shape and the permeability of the flux path. In a transformer the main flux is contained within an iron core, where 'L' is the flux path length of the core, and 'a' is the core area. The core is made from a material which has a high permeability 'μ' and the reluctance is given by;

$S = L/\mu a$

The flux density (B) is amount of flux per unit area.

$B = \Phi/a = I/Sa = \mu I/L$

Due to the high permeability of the iron, a small current can achieve the flux required. This is known as the magnetizing current. There are usually a number of turns 'N' associated with a power transformer, and together they form the magnetizing or primary winding. This winding is wound around an iron core, which forms a closed loop through the winding and the main flux is contained within this iron core.

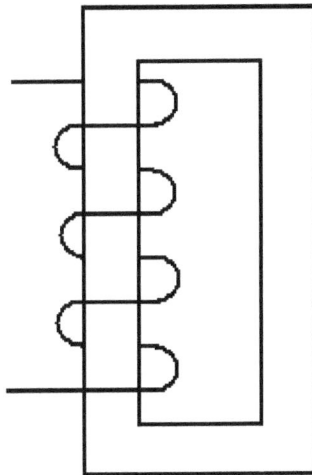

Figure – 2

In Figure – 2, the flux density is a function of the ampere-turns and is given by;

$B = \mu IN/L$

If a sinusoidal voltage is applied to this winding which has a frequency 'f' cycles per second, then a sinusoidal flux of the same frequency will exist in the core.

If there is a secondary winding also wound around the core, then a voltage will be induced in this winding. This voltage will follow the rate of change of flux. $E = d\Phi/dt$

If the flux is sinusoidal and changing from $+\Phi_m$ to $-\Phi_m$ in half a cycle then for a frequency of f cycles per second the average rate of change is $2\ \Phi_m\ /\ 1/(2f)$.

The average voltage is therefore $E = 4f\ \Phi_m$, The more common value to use is the RMS value which, for a sinusoidal wave is 1.11 times the average value therefore the volts for each conductor (Turn) is given by;

$E = 4.44f\ \Phi_m$ volts per turn

If there are N turns then the voltage is $E = 4.44\ N\ f\ \Phi_m$

Transformers, as the name implies, transform the voltage levels in one system to a new voltage level in another system. This is achieved by inducing a magnetic flux in an iron core using one winding and linking this flux to another winding. The voltages in the windings will be proportional to the number of turns in each winding, the flux linkage can therefore be used to increase the voltage or decrease the voltage. In the following example, the first or primary winding has fewer turns than the secondary winding and the voltage is therefore stepped up.

If a load is placed on the secondary winding then a current will flow. This current will create a flux around the windings, and this is known as the leakage flux. If the coils are closely coupled, the inductance of the coils will present lower impedance to the supply and a current will flow in the primary winding. This current will be sufficient to counteract the amperes in the secondary. As there will be different turns in each winding the ampere turns will balance. The energy is transferred through the gap between the windings and is only restricted by the load and the inductance of the windings.

For a standard core type transformer, there are at least two windings. The first winding or low voltage winding (LV) has a small number of turns with a large cross sectional area of copper to carry a high current. The second winding or high voltage winding (HV) has a large number of turns and a small copper area, which carries a small current.
The relationship apart from losses, is that the LV current x LV voltage is equal to the HV current x HV voltage; it is nearly 100% efficient. So that the power input is equal to the power output.

Normally large power transformers are built as three-phase units and the nature of the three phases which are 120° apart allow the flux in each phase to be returned through the other two phases. The most efficient use of copper is to connect the high current winding in a delta and connect the high voltage winding as a star. Figure-3 shows the arrangement graphically, where the primary winding is termed the LV and the secondary winding is termed the HV. It should be noted that 'you cannot get something for nothing' and the increase in voltage is associated with a decrease in current.

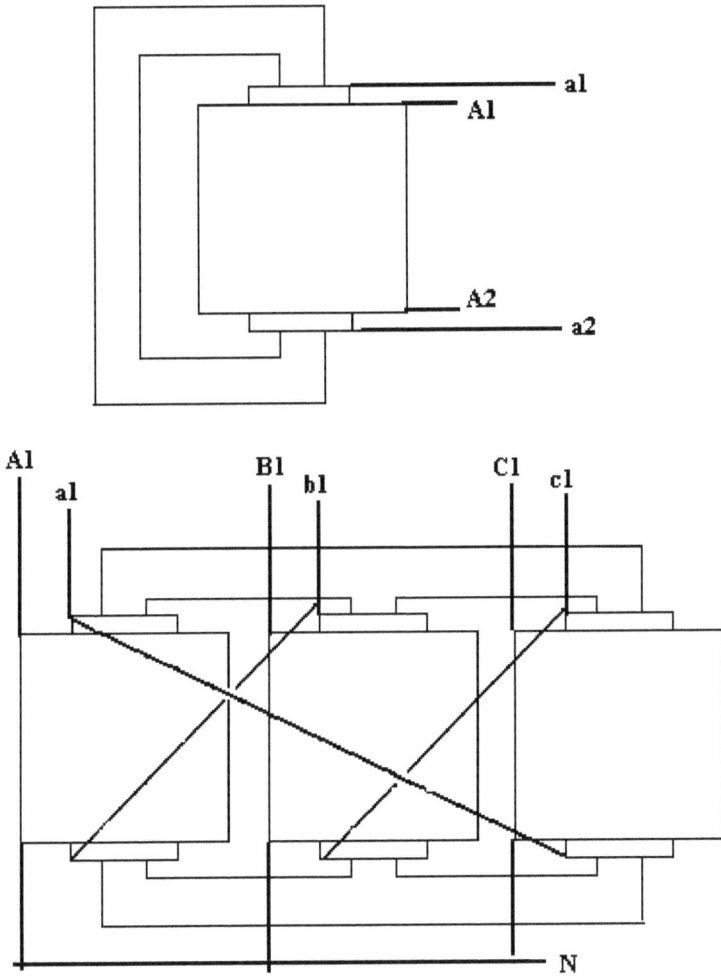

Figure-3

The fundamental operation of all transformers depends on the theory that all magnetic fields operate with a corresponding electric field at right angles to it. By using a ferrite material (the iron core) to concentrate the magnetic field and a corresponding conductive material (the copper winding) to concentrate the electric field, then energy may be transferred through the coupled circuit. If the fields are controlled then this energy transfer may be very efficient.

There are two main modes of operation, core type and shell type as shown and both are as efficient. The energy may be transferred in single or three phase mode as shown.

There are many arrangements for cores and windings as shown in Figure-4,

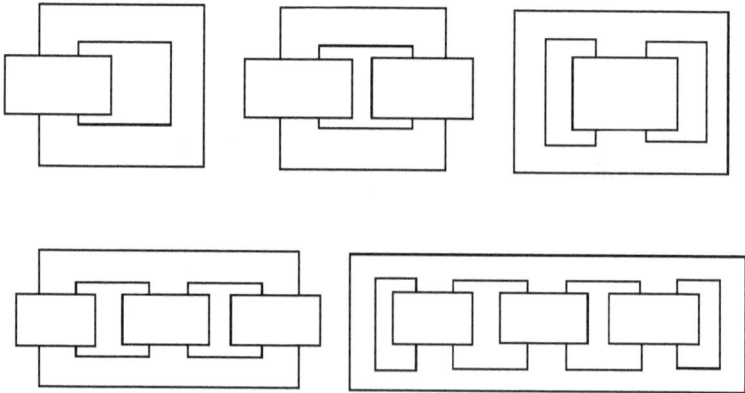

Figure -4

Flux Density

The cross sectional area of the core, which is normally manufactured from flat steel plates with mitered joints, is not uniform and particularly in the corner areas it has a much higher area therefore the flux density throughout the core will vary. The pattern is even more complex for three-phase transformers where there is 120° displacement between phases. The term flux density refers only to the maximum value, which is taken to be the cross sectional area in the middle of the leg where the area is a minimum and the flux lines are parallel to the leg length. The core plates are assembled with alternate layers providing an overlap at the corners; this is shown as an example in Figure-5. There are several ways to create cores using flat plates and often a step-lap arrangement is used. In this case there may be 5 or seven smaller steps in each packet. The little projection at each corner may be 25-30 mm. Each core plate is insulated with a coating that prevents circulating current between the plates. It may also be advisable to provide extra insulation between large sections of core with a sheet of paper or NOMEX. The plates are all held together with glass fiber cured bands: there may be some smaller cores that still use core bolts but these are not recommended.

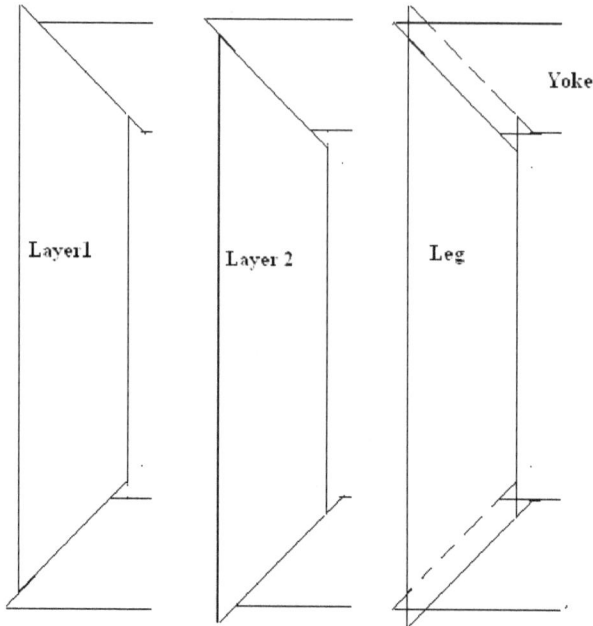

Figure-5.

The losses in the core are measured at full voltage and rated frequency under no-load conditions. They are often measured at 90%, 100% and 110% voltage levels in order to assess the degree of saturation or how near the operating flux density is to the 'knee' point. There are many types of cores and there are various modes of operation, depending on the type of regulation required. The cores may be operated at constant flux, in which case the turns ratio is used to change the voltages, for fixed ratio step-up transformers the generator voltage may be increased or decreased to control the output voltage, the unit may have neutral end taps which can be used to vary the HV or the LV in an autotransformer; in which case the flux density will vary. Some units vary the turn's ratio and the flux in order to match a variable supply and variable load to maintain a fixed load voltage. The design of the core should take into account the most onerous flux condition and not only the maximum loss but also the temperature at the hottest point in the core. Cores are usually designed with cooling ducts located to give the optimum cooling. It is known that gases can be generated in hot cores and they can lead to insulation failure if they form bubbles, which can migrate into the insulation structure.

Magnetic Characteristics

The characteristic of the magnetizing impedance is non-linear and depends on the properties of the core steel, the type of core, and the applied voltage. In most cases, due to the large permeability of the steel, the total flux generated by the voltage across the winding is confined to the core and this total flux will produce a flux density in the core according to the cross sectional area, provided that the core does not saturate. The effect of saturation is to reduce the permeability of the steel and therefore increase the magnetizing current. A core steel characteristic (not to scale) is shown in Figure-6, to demonstrate an over-flux condition.

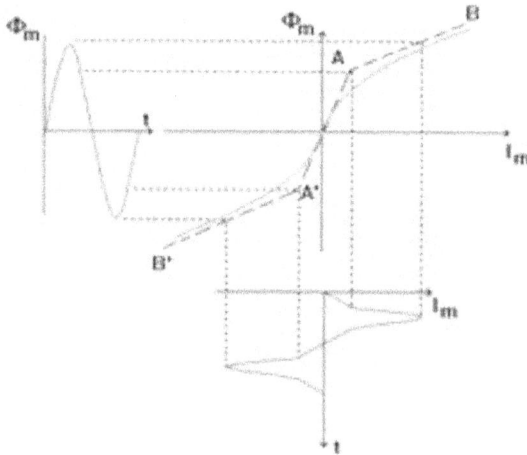

Figure-6

A sinusoidal voltage is applied which creates a flux in the core. The characteristics for the core steel are shown as a continuous line, the asymptotes to this line are indicated by B′A′AB. It will be noted that the magnetizing current will be 'peaky' where the saturation occurs.

In cases where there may be some initial retention of flux in the core, prior to excitation, this is termed Remanence, there will be an initial DC component of flux which will add to the AC flux as shown in Figure-7. This will cause a lop sided magnetizing current which can lead to an initial 'in-rush' current, and this may have very high levels that may last for several minutes for a large core. By energizing the core for a short period of time, the DC component will reduce to zero. This can often be observed by the reduction in noise level after this initial period.

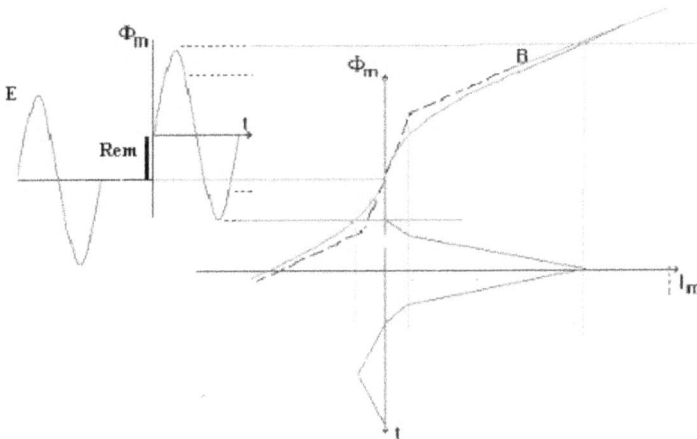

Figure-7

Modern measuring instruments can analyze the magnetizing current instantaneously and by observing the odd harmonics, it can be seen that they reduce within a few minutes, thus removing the effect of the remanence.

The noise generated in a core which is energized, is associated with magnetostriction. This is the phenomenon whereby there is a dimensional change in the core plate when it is subjected to a magnetic field and this movement creates noise. The magnetostriction is a non-linear function of the flux density and a typical curve is shown.

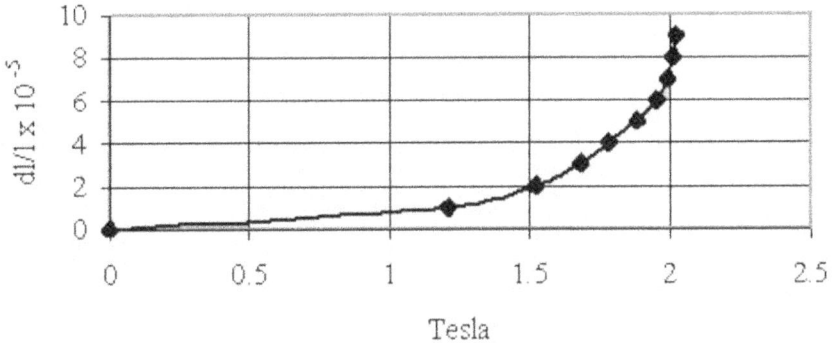

Figure-8

The core construction may be represented by components as shown in Figure-9

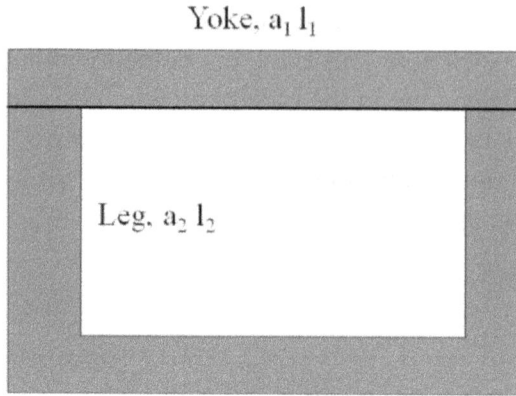

Yoke, a_1 l_1

Leg, a_2 l_2

Figure-9

This core may be modelled as a mass and a spring, as shown in Figure-10. And the resonant frequency of this model (f_o) may be estimated by considering the mechanical resonance. The top yoke is considered as a mass M sitting on the leg, which is considered as a spring with a stiffness factor s.

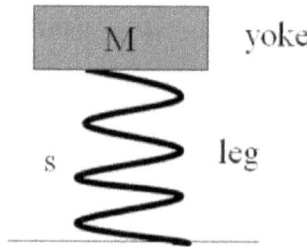

M   yoke

s   leg

Figure-10

$$f_0 = \sqrt{\frac{s \times g}{M}} \div (2 \times \pi)$$

where

$$M = \frac{l_1 \times a_1 \times d}{2}$$

$$s = \frac{2 \times E \times a_2}{l_2}$$

If this mechanical frequency is close to the electrical supply frequency then vibrations at this frequency may be enhanced.

To prevent excess noise, it is advisable to operate the core below the 'knee point' when the transformer is operating at normal flux. The core should not exceed the required temperature rise when operating at the maximum value: this is the level which will dictate the number of cooling ducts.

Windings

A preliminary design which selects the core size, flux density and the winding dimensions, will also calculate a value for the volts per turn. It is then up to the design engineer to check that this value is allowed. There may be an exception in special cases where half turns may be used, but normally all the windings required to have an integer number of turns. In High ratio transformers with small tapping sections, the smallest tap, which may only be 1% of the winding will dictate the allowable integer number of turns, and hence limit the volts per turn to some discreet levels. Some juggling of core size and winding lengths, may be required to accommodate the turns ratios available which provide, within the required tolerance, the voltage ratios specified.

The windings in a core-type transformer are normally round concentric coils, wound around the core with the lowest voltage winding on the inside and the higher voltages outside. This arrangement gives the most efficient use of space and the easiest insulation structure. There are however other combinations, such as double concentric windings, which have over riding advantages.

In the initial design of a transformer it is usual to define a required space for the winding including the internal insulation and cooling. This winding 'block' will include the conductors, and the ratio of conductor area to the total area can be termed the winding space factor. The choice of winding length and diameters in conjunction with the core diameter and space between the other windings will dictate the required transformer reactance, which will be fixed by the specification.

The winding block will require space for the conductor, the insulation and the cooling, and these factors will depend on the voltage levels and temperature rises. Detailed dimensions of the conductor, insulation and cooling ducts form the design of each winding, and there are many variations available which will suit the style of the transformer. A typical cross section is shown in Figure-11, which utilizes a simple papered-covered strip conductor with radial cooling ducts. The cross sectional area of each turn is determined by the rated current of the winding and the chosen current density. It is normal in large transformers to sub-divide the conductor area into a number of parallel strips, which reduce circulating currents within the conductor. Transpositions between the strips are required to ensure that each parallel strip has the same resistance and reactance and therefore carries an equal share of the current.

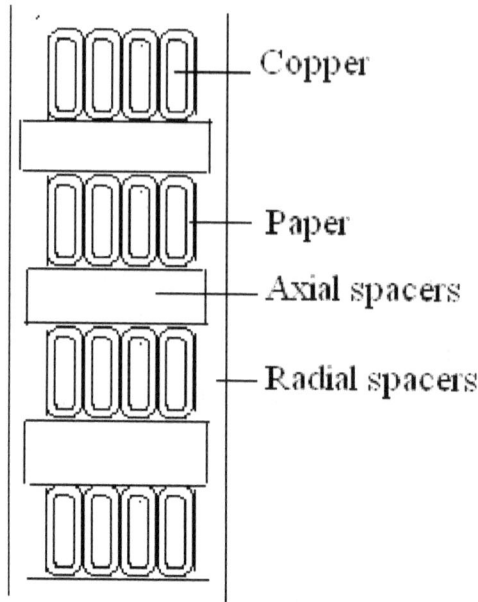

Figure-11

The thickness of paper required for each conductor will depend on either the operating voltage or the test voltage, whichever is the more onerous.

The axial spacers, or stampings are keyed on to the radial spacers or sticks, in such a fashion that they can slide when the coil is compressed. Major insulation suppliers can provide this type of arrangement as a kit. The number of stampings per circle will be determined by the cooling requirements and the mechanical forces generate during fault conditions. It is normal to impose a limit on the stampings of around 35 MPa, and precompressed pressboard is usually required.

The radial spacers, or sticks are required to provide a cooling-oil flow path for the length of the winding: they are distributed equally around the circumference of the winding, taking up around 30% of the conductor surface, the other 70% is required for the cooling oil.
The copper used may be supplied in several options, and the manufacturers will offer single strip, twin strip, triple strip or transposed cable both bonded and unbonded. The subdivision required will be controlled by the percentage of stray losses that are acceptable in the winding. The individual strips within the transposed cable are normally insulated by using enamel. If added strength is required these strips may have an epoxy resin coating, which when cured, bonds the strips together thus increasing the overall strength of the cable.

There are several styles of winding available, ranging from simple spiral windings to complex hybrid windings using interleaving sections and continuous disc sections with continuously transposed cable. The choice of style will depend on the operating currents and voltages, and the tests required: account must also be taken of the final position of the leads, there is an option to take the leads from the top and bottom of the winding or the leads on the outer winding may be taken from a center point with the winding in two halves. This arrangement is very often used with very high voltage terminals as the insulation structure is less complex. Windings carrying high currents will require that the conductor is subdivided in to a number of smaller parallel conductors to avoid excessive stray loss, each small conductor strip will require to be insulated from the other strips. Care must be taken when using arrangements which involve parallel conductors: suitable transpositions are required to avoid circulating currents between the parallel paths. The manufacturers of conductors can provide multiple strip cable, where the insulation is enamel and the individual strips are continuously transposed, the cable can also have an overall paper covering. Very high current windings may have several of these cables in parallel and transpositions will be required between the cables. Simple spiral windings will have a helix and require a wedge type support to present the supporting structure with a flat clamping surface. Spiral windings are also used where an internal tap winding with several sections is required. In this case the winding sections are all wound in parallel using a multi-start arrangement, they are then connected in series using external leads, often known as 'organ pipes' due to the similarity of construction. An example of the interconnections is shown in Figure-12.

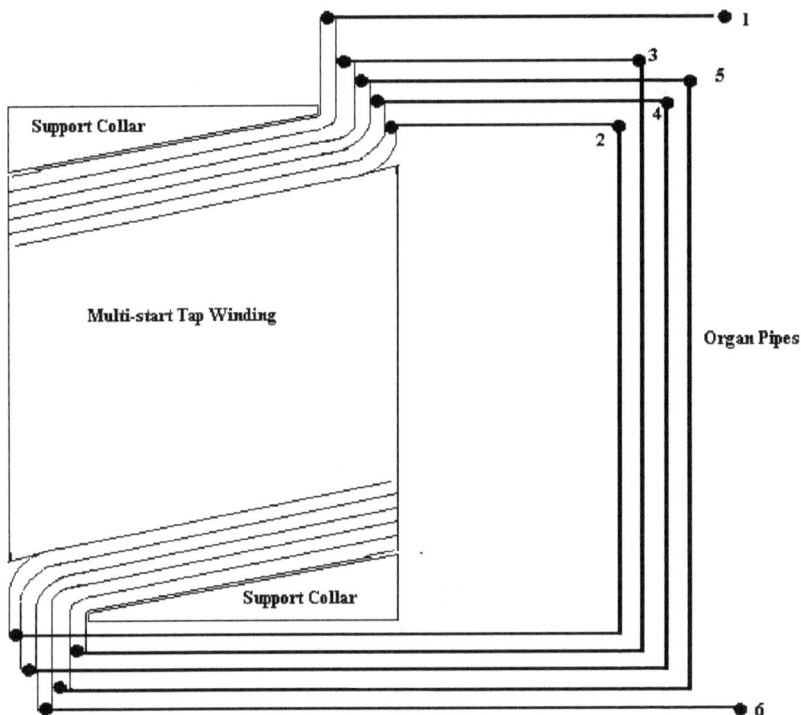

Figure-12

Interconnecting in this fashion will restrict the operating voltage across the conductor insulation to no more than twice the section voltage. This arrangement will also have a high series capacitance as the voltage between adjacent conductors is that of two sections, and this will help the initial capacitive distribution during impulses.

The continuous disc winding is very popular for large power transformer windings: it offers the designer options for radial cooling ducts and high series capacitance. As the winding is continuous, without any breaks or connections, is can employ continuously transposed cable. The arrangement of the turns for one group is shown in Figure-13 and as there is a facility to introduce transpositions, parallel conductors do not present a problem. It is more convenient to wind this type of winding on a vertical lathe as this eases the disc support during manufacture.

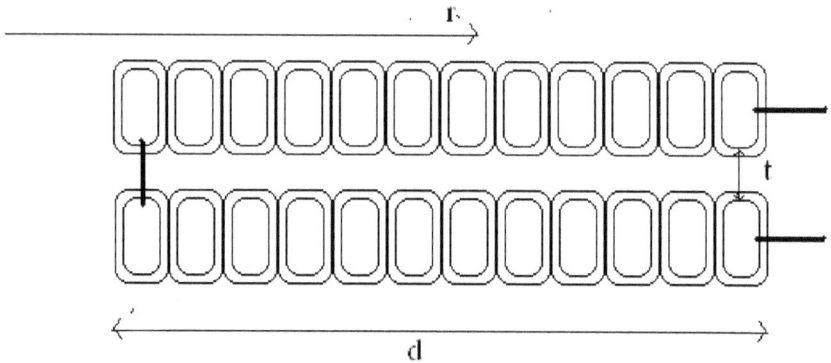

Figure-13

r is the mean radius of the section
t is the distance between the conductors
d is the radial dimension of the winding
N is the turns per group
V is the volts per turn and the group voltage is N times V.

The insulation required between the turns is paper and the thickness of the paper is controlled by V. The insulation between sections in the group is controlled by N times V and the initial distribution when impulse voltages are applied to the winding. The insulation between the sections may consist of a composite arrangement of paper, oil and pressboard.

In order to estimate the initial impulse distribution, it is required to estimate the series and shunt capacitances of the group.

One technique is to break the group into bits and estimate the capacitances of the bits, then equate this to an equivalent group capacitance by summing the stored energy.

The capacitance between two adjacent conductors, as shown in Figure-14, may be estimated by considering the conductor dimensions and material permittivity.

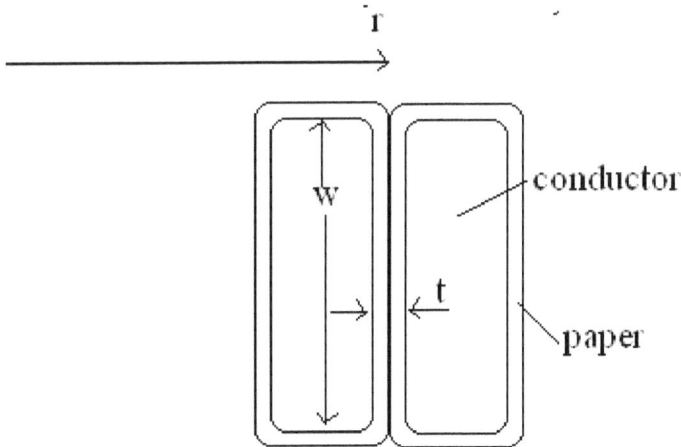

Figure-14

$Ct = 2\pi r(w + t).\varepsilon_0.\varepsilon_r / t$

Where;

C_t = Capacitance between turns
$2\pi r$ = Length of mean turn of group
W+t = effective width of conductor including fringing
ε_0 = permittivity of free space 8.85 x 10^{-12}
ε_r = relative permittivity of paper = 3.5

The energy stored per turn = $C_t.V^2/2$
The energy stored between turns in the group with N turns
= $(N). C_t.V^2/2$

The capacitance between the two sections in the group is given by
$Cs = 2\pi r.d.\varepsilon_0.\varepsilon_r / t$

In this case the distance between the conductors may consist of a composite dielectric containing paper, oil and pressboard, which have different relative permittivity.
 The term ε_r / t may therefore be replaced with a composite distance which reflects this different relative permittivity.

$T = (t_1/\varepsilon_1 + t_2/\varepsilon_2 + t_3/\varepsilon_3)$

The energy stored between sections = $Cs.(NV/2)^2 / 2$

There will also be a similar capacitance between the groups, although the spacing may be slightly different.

$$C_g = 2\pi r.d.\varepsilon_0.\varepsilon_r \, / \, t$$

The energy stored between groups $= C_g.(NV/2)^2 \, / \, 2$

The sum of the energy stored for the group equates to an effective capacitance with an applied group voltage NV is given by;

$$C_{eff} = (C_t/N + C_s/4 + C_g/4)$$

If the turns in the two sections are interleaved as shown in Figure-15, then the voltage between the adjacent conductors is NV/2. The energy stored in the capacitor associated with the turns is therefore;

Energy stored due to inter-turn capacity = (N). $C_t.(NV/2)^2/2$ Which is $(N/2)^2$ times the energy stored in the same group connected as a continuous disc. This improves the overall series capacitance significantly and hence reduces the α factor in the calculation for the initial impulse distribution. The disadvantage being that there will be more conductor paper required and the voltage between the groups will be 1.5 times the voltage between the sections.

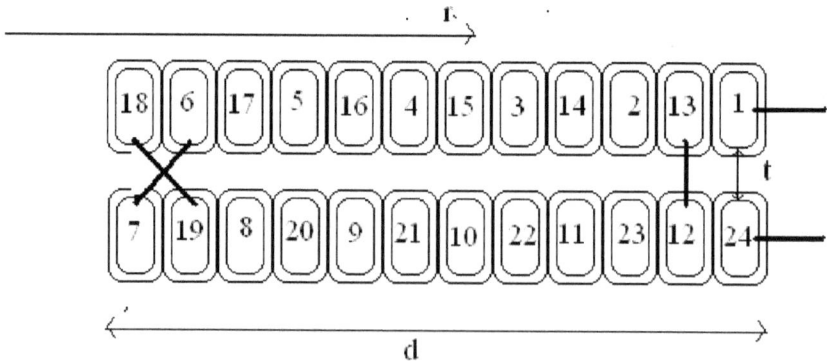

Figure-15

Due to the complexity of the interleaved group, it is common to restrict this type of winding to a few groups at the line end, and use continuous disc for the remaining groups.

Winding Leakage Flux due to load current

As shown previously in Figure-1, a magnetic field (Φ) is produced around a conductor when an electric current (I) flows through this conductor. The relationship between the strength of this field and the magnitude of the current is the reluctance (S) of the path of the magnetic field.

The leakage flux due to the load current flows in the windings and the surrounding medium, which is predominately non-magnetic. The ampere-turns (AT) is the current times the number of conductors carrying this current.

The reluctance is a function of the shape and the permeability of the flux path and is given by;

$$S = \text{length(l)/area(a)} \times \text{permeability}(\mu)$$

The flux density (B) is amount of flux per unit area.

$$B = AT/aS \quad = AT/\mu l$$

Each winding will therefore have a flux associated with the winding, which will be a function of the current in the winding. This flux is different from the core flux as it is a function of the current and mainly flows in the high reluctance path around the winding and in the gaps between the windings. It is known as leakage flux and is represented in the equivalent circuit as an inductive reactance through which the load current flows. This reactance is in series with a resistor, which represents the loss in the windings. As the load current increases, the leakage flux will increase and the current will be drawn from the source to balance the ampere turns. Thus the energy will be transferred through the transformer leakage reactance.

Note also that due to the large non-magnetic content, the series branch representing the windings has a linear characteristic as shown in Figure-16.

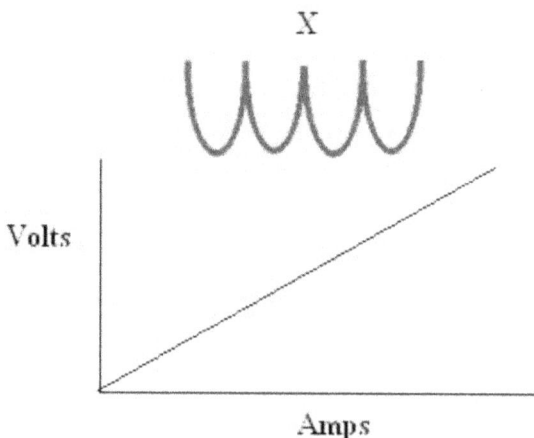

Figure-16

The reactance X is formed by a series of inductances which will be the sum of various self and mutual inductances operating at a fixed frequency.

Consider the two concentric coils wound around a steel core as shown in Figure-17.

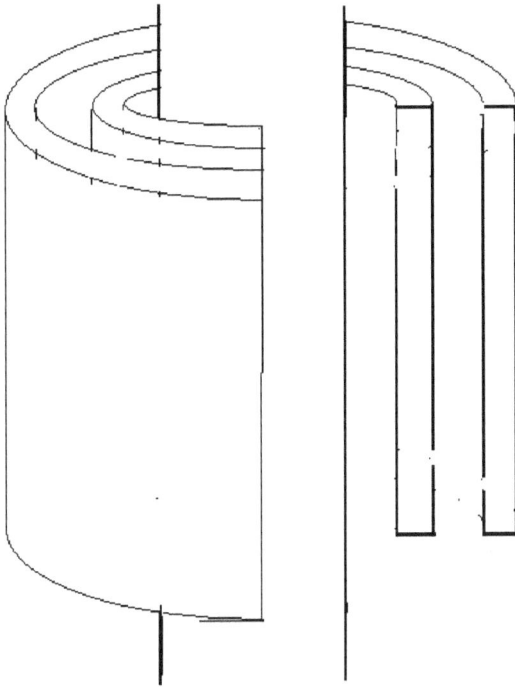

Figure-17

There is a combination of inductances made up from the self-inductance of the inner coil, the self-inductance of the outer coil and the mutual inductance between the coils. Consider now a section of one half through the centre which when rotated through one hundred and eighty degrees will represent the complete coils.

This in turn may be drawn as two inductances with a mutual inductance as shown in Figure-18. Neglecting the coil resistances and assuming a frequency of ω radians per second, two

45

equations may be formed to determine the series reactance for a short circuit secondary.

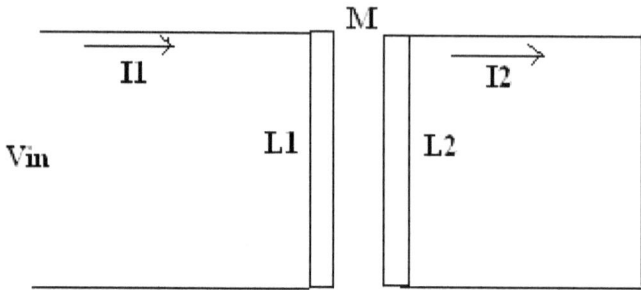

Figure-18

$Vin = I1.j\omega L1 + I2.j\omega M$

$I1.j\omega M = I2.j\omega L2$

Solving these will give the input reactance at the primary as

$\underline{Xin = j\omega.(L1-M^2/L2)}$

Reactance

The reactance of a transformer is a measure of the amount of leakage field within the transformer under full load conditions. The main part of the leakage field flows in the gaps between the windings, where the maximum number of ampere-turns is required to circulate the flux, due to the fact that there are no ferro-magnetic materials in these regions and the relative permeability is unity. The return path for the flux, is usually via the route taken through the core, for the inner windings and through the tank or tank shields for the outer winding, the return path has a much higher relative permeability and therefore has much less influence on the required ampere-turns.

In order to make an initial assessment of the leakage flux in the gaps, the effect of the core and tank is neglected and the flux in the gap will be directly proportional to the ampere-turns.

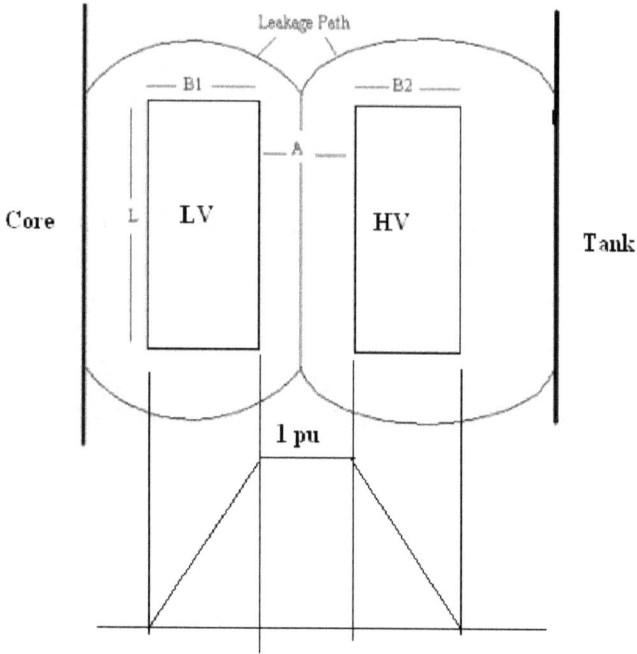

Figure-19

Consider Figure-19, a core type transformer with concentric windings as shown, then starting from the inner surface of the inner winding and proceeding to the outer surface of the outer winding. The available ampere-turns will form a pattern as shown, where there are zero ampere-turns inside the inner winding; this value will approach unity as we proceed to the outer diameter of the inner winding, and remain at unity across the gap. The balancing ampere-turns in the outer winding will start to take effect and the ampere-turns will reduce to zero at the outside of the outer winding.

Due to the constant permeability, the flux density will follow the same pattern, and the total leakage flux may be estimated. The transformer reactance is often represented as the ratio of the total leakage flux due to full load current to the actual total no-load flux in the core.

The transformer leakage flux pattern is a function of the shape of the transformer. It may be short and fat or long and thin. The reactance of course is a specified value and must be the same for any given shape. The relationship between the various dimensions can be studied to estimate the best design to offer based on other specification limitations. It is useful to look at the simple equation for reactance to determine these relationships.

Dimensionally the Reactance is –
Proportional to the winding radial dimensions, the winding gaps and the ampere-turns, and inversely proportional to the winding height, the core diameter and the volts per turn.

For example if there is a tight shipping height, then the design may have to be short and fat, in which case a larger-than-normal core diameter may be required. If the unit has a shipping weight restriction then it may require a smaller core diameter with a higher-than-normal winding height. See Figure-20.

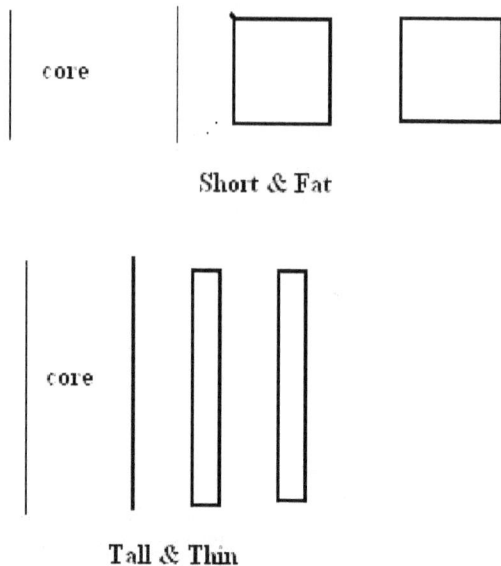

Short & Fat

Tall & Thin

Figure-20

Note – apart from the shape, these two transformers will meet the same specification.

Reactance can usually only be measured by supplying one circuit and shorting another circuit in order that there is a balance of ampere-turns. Either or both circuits may have more than one winding in series. It these cases, is usual to use a computer program to determine the complex field, however, the reactance may be estimated by using a simple superposition technique whereby each reactance pair can be calculated by a simple formula, and the pairs combined to determine the total reactance between the two circuits. In the examples shown, the primary circuit winding are shown in black and the secondary circuit windings are shown in grey. The corresponding ampere-turn diagram is also shown.

Example-1 – Consider a two-circuit transformer with a single LV and a separate tap winding in the HV. As shown in Figure-21.

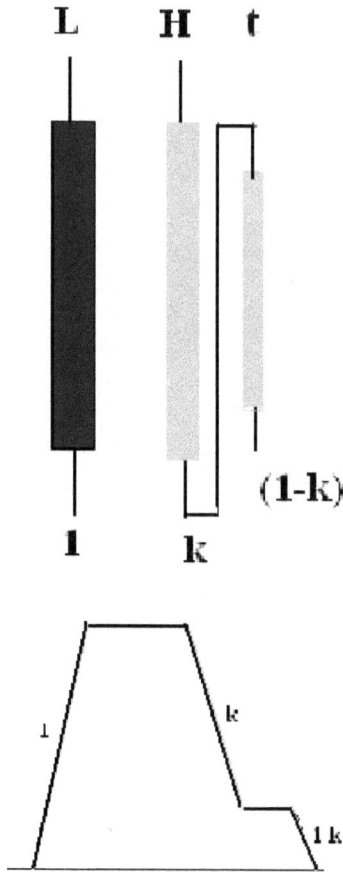

Figure-21

The reactance between the two circuits is given by:

$$X = Xlh(1)(k) + Xlt(1)(1\text{-}k) - Xht(k)(1\text{-}k)$$

Where all the reactance components are calculated on the same base.

Example-2 Consider two circuits with a number of individual windings in each circuit as shown in Figure-22, one circuit is black and the other circuit is grey.

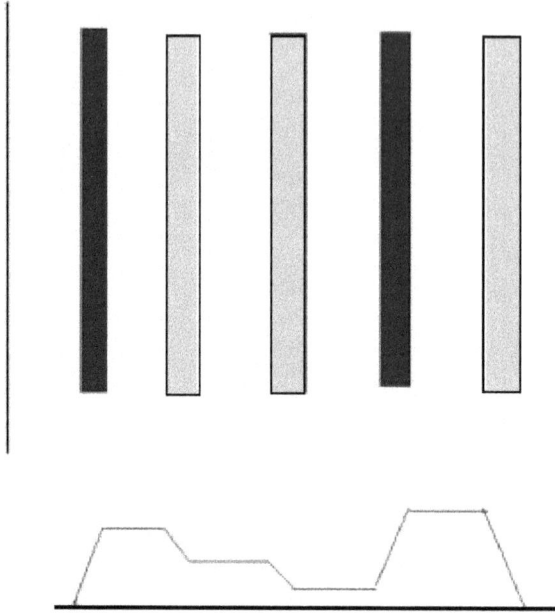

Figure-22

In this case, numbering the windings from the core out, the total reactance between the two circuits is given by:

$X = X12(k1)(k2) + X13(k1)(k3) - X14(k1)(k4) + X15(k1)(k5)$
$\quad - X23(k2)(k3) + X24(k2)(k4) - X25(k2)(k5)$
$\quad + X34(k3)(k4) - X35(k3)(k5)$
$\quad + X45(k4)(k5)$

Note that the shape of the ampere-turn diagram, or the area under the curve, is proportional to the amount of leakage flux and hence the value of reactance.

In the case of a tap winding where the direction and quantity of ampere-turns can be controlled. The reactance can be made to increase with tap, or decrease with tap as required. This is achieved by placing the tap winding in different physical positions as shown in Figures-23a, 23b and 23c; the tap winding is white since it can be associated with the LV or the HV depending on the B-B connection. The vertical axis indicates the % change in the impedance depending on where the tap winding is located.

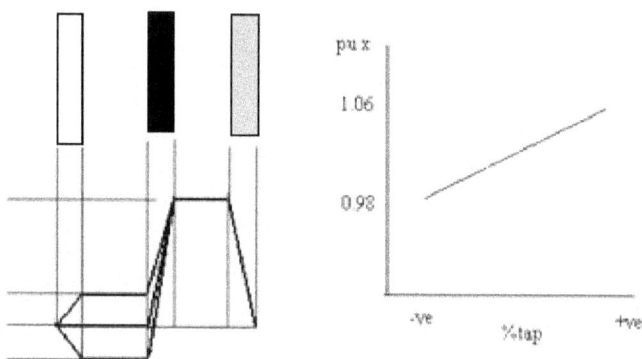

Figure-23a

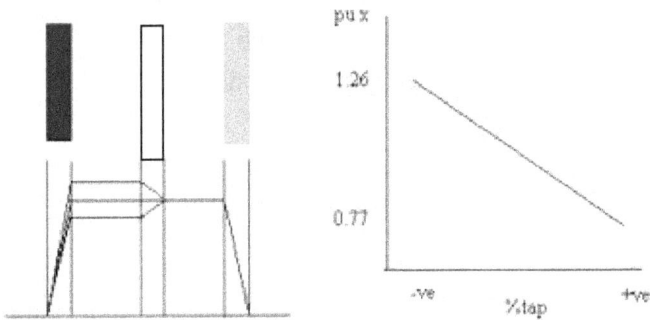

Figure-23-b

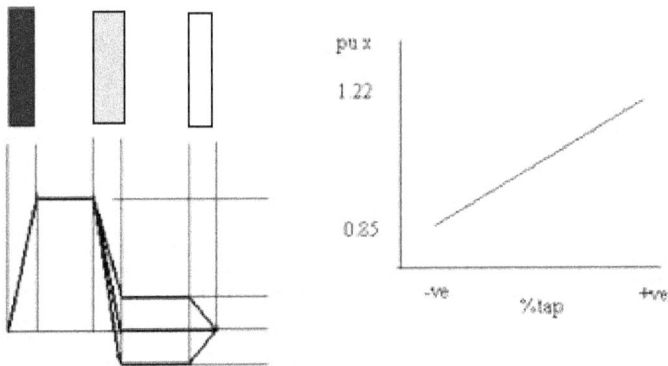

Figure-23c

The analysis of the transformer however, is usually carried out on the single phase equivalent circuit as shown.

There are two paths for the magnetic field (or flux), one is the path that is restricted to the core, and is the main exciting flux.

The other is the flux that flows outside the core for most of its length and this is the leakage flux. It is the leakage flux that dictated the 'Impedance' of the transformer as this flux is proportional to the load current in the transformer.

The main flux which is restricted to the core is represented by a parallel or shunt branch and has a non-linear characteristic controlled by the type of steel used. The leakage flux is represented by a series branch and as the flux path is in a non-iron path, the characteristic is linear and is a function of the load current in the transformer.

These main and leakage fluxes and the winding resistances and core losses may be represented by combining the effects as shown in the equivalent circuit. This circuit may then be used to analyze the performance of the transformer at various voltages and currents.

When the load is added to the equivalent circuit, the leakage of the LV and HV may be combined to give the above simplified equivalent circuit for easier analysis, as shown in Figure-24.

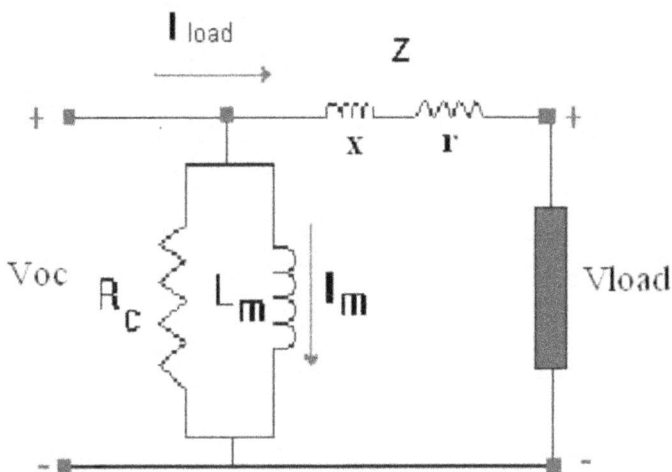

Figure-24

The operation of the transformer can then be predicted by solving the equivalent circuit for various voltages and loads. A typical phasor diagram is shown with the various voltages and currents.

Voc = Open circuit voltage,
Vload = Voltage under load,
I = Current
x = % reactance,
r = % resistance,
φ = load angle

It can be seen that the voltage under load is smaller than the no load voltage for a lagging power factor load as shown in Figure-25.

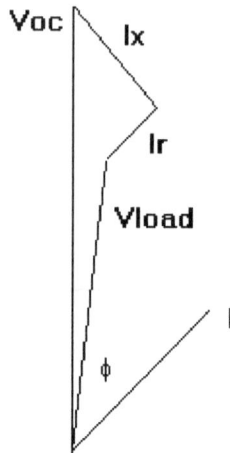

Figure-25

The voltage drop due to the load is called the regulation and it is normally represented as a percentage of the no load voltage. This may be counter acted by either increasing the output voltage, by the use of extra turns in the output winding, or increasing the volts per turn. The former method is described as constant flux regulation and the latter by variable flux regulation. If it is required to alter the flux for this reason, it will require that the core is able to operate at the most onerous flux density and normally a larger core is required.

The regulation required will determine the number of extra turns required, or the 'tapping range'. Adding or subtracting turns from a winding or circuit will have an effect on the value of the reactance and this must be taken into account.

The Flux due to the load current in the windings is termed the leakage flux and mainly flows in the windings and the surrounding oil. A typical leakage field plot is shown in Figure-26. It can be seen that the leakage flux is mainly axial and is the full length of the windings with some radial near to the winding ends. This radial component will find a return path, mainly in the core for the LV and in the tank for the HV. The quantity and distribution of this leakage flux will determine the transformer reactance and it is through this reactance that the load current flows and transfers power from one circuit to the other. The leakage flux patterns will vary significantly for different winding designs and there may be up to 5 loaded windings in a large transformer. It is important that the windings are magnetically balanced as these leakage fields can be very high under fault conditions, and in conjunction with the fault current they can generate very high forces.

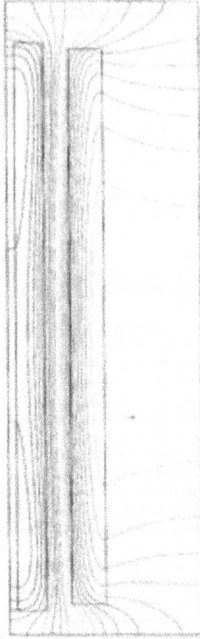

Figure-26

In addition to the normal I^2R losses in the conductors, the leakage field will react with the conductors and generate local circulating currents that will add to the load loss. This additional loss will not be uniform throughout the windings and there will be local hot spots generated.

As can be seen from the equivalent circuit, shown in Figure-25, when the load changes, the internal impedance voltage will change and the output voltage will change. As this may not desirable, it is common practice to compensate by introducing more turns into the output winding, these are known as tapping turns and they are switched in and out by using a Tapchangers. The load variation can be predicted from the daily load profile as shown in Figure-27

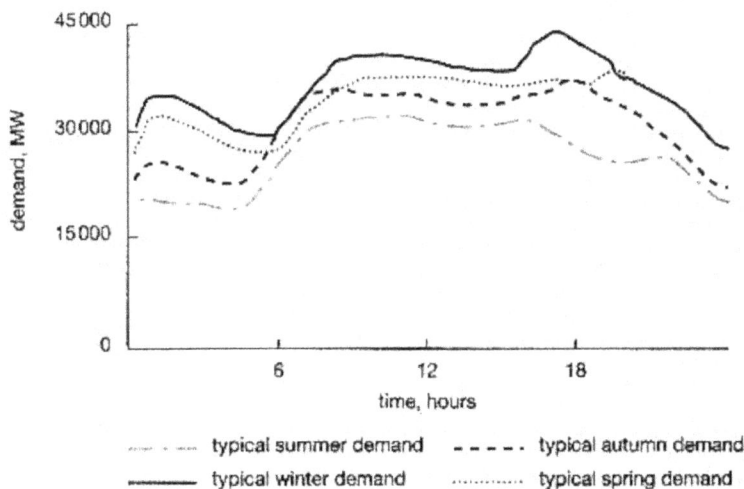

Figure-27

The appropriate tapping range can then be estimated to keep the output voltage within the specified limits.

Tap Changers

The tapping range chosen will have to compensate for the internal voltage drop or regulation and also some variation in the line or system voltage which may change. There is also a variation due to the change in impedance of the transformer when the tap winding is introduced. Regulation is also a function of the load power factor and this can have a major effect on the required tapping range, it should be agreed with the customer before the design is completed.

There are various type of tap winding and various options for electrical and physical placement. There are three options shown in Figure-28, two for autotransformers and one for a generator transformer. The neutral end taps in the auto-transformer will create a variable flux condition and may be used for HV or LV variation; it will also vary the tertiary voltage. The other two connections will maintain a constant flux in the core.

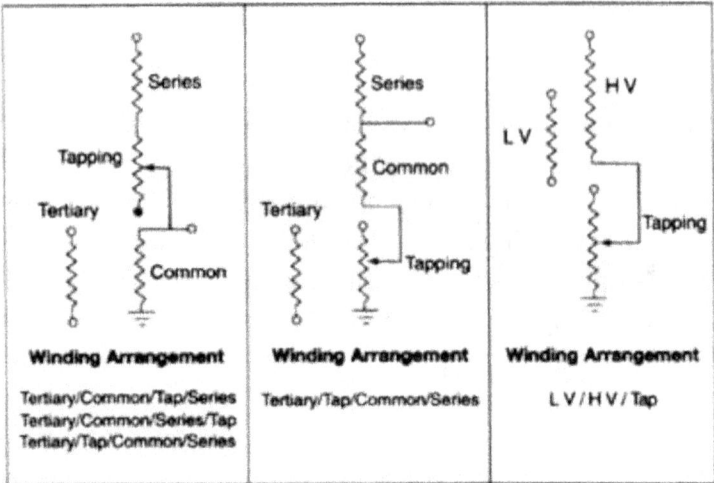

Figure-28

Figure-29

Figure-29 shows a typical arrangement of HV LTC's located at the line end of a large autotransformer.

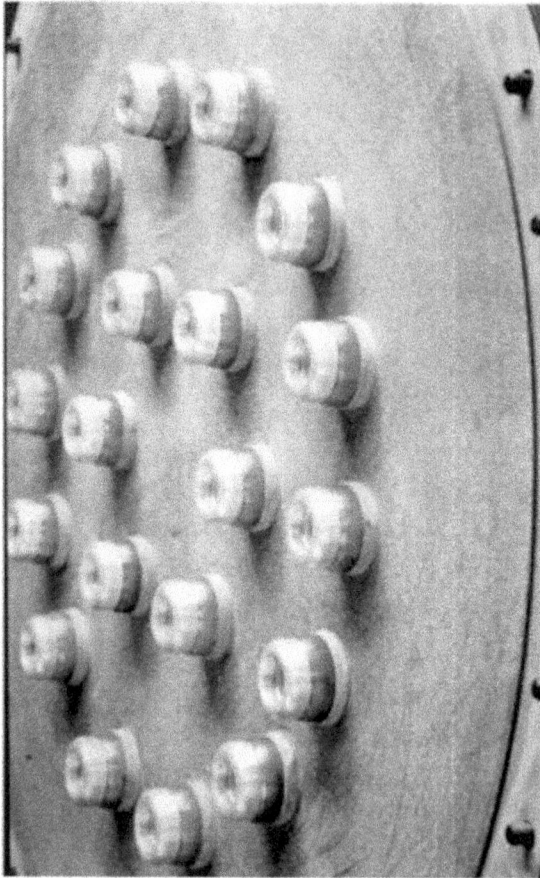

Figure-30

Fogure-30 shows a sample barrier board, which separates the oil from the main transformer. This allows the tapchangers to be removed for shipping purposes and keeps the oil separate during normal operation.

Figure-31

Figure-31 shows a large neutral end Tapchangers as used in GT's with a section for metal oxide varistors, which protect the tap changer against high voltage surges.

Figure -32 shows various mounting arrangements for OLTC's.

| bolt-on type | in-tank with barrier board | in-tank common oil | weir type |

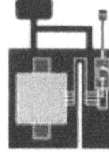

Figure-32

Thermal Considerations

Operating temperature are mainly represented as temperature rises above ambient, this applies to the winding and the oil. The main reason for fixing the temperature is to theoretically fix some boundary for the lifespan of the transformer. The main insulation in oil filled transformers is the paper surrounding the copper. This paper is usually a cellulose base and as such it is made from fibres formed by cellulose chains, which give it mechanical strength. These chains will break over a period of time and the rate of the number of breaks is a function of time and temperature. The measure is termed the degree of polymerisation (dp) and the insulation is deemed to have a very low mechanical strength when the dp falls below 300, 1200 is new insulation. In order to have an expected life of 30 years, the hottest temperature of the insulation should average less than 100 °C over this period of time. This may involve periods below 100 and periods above 100 °C but there are limiting factors which restrict the upper level to about 120 °C, at 130 °C bubbles can form and the local dielectric constant changes and breakdown can occur. Depending on the ambient conditions the average winding rises can vary between 50 °C and 70 °C. There are obviously various conductor materials and cooling mediums, which allow higher temperatures and they will be discussed later.

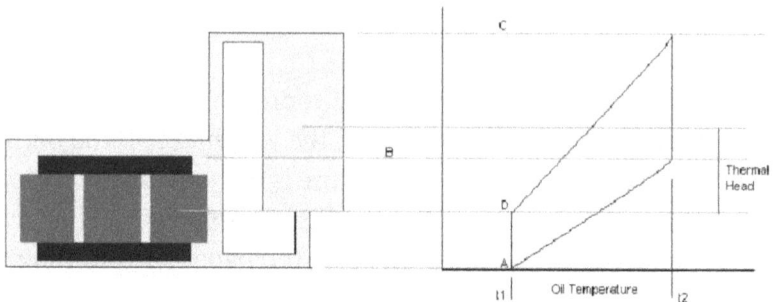

Figure-33

The arrangement for the oil flow in an oil cooled transformer is shown in Figure-33, where the temperatures at various positions are indicated on the graph. The bottom oil entering the tank is t1 at position A; the top oil temperature leaving the tank is t2 at B The thermal head is the difference between the average height of the effective coolers and the average height of the windings. This thermal head is essential to provide a driving force to circulate the oil under ON conditions, and it is normally greater than 0.5 m.

The transfer of heat from the winding copper to the oil through the paper is dependent on the temperature difference between the copper and the oil, which has two components: A gradient across the paper, (t3-t2) which is dependent on the thickness of the paper, and a surface heat transfer between the paper and the oil, (t2-t1) which is dependent on the flow of the oil. There will also be a surface heat transfer between the oil and the air through the radiator surface, (t1-t0) which will be dependent on the air flow. This is shown in Figure-34.

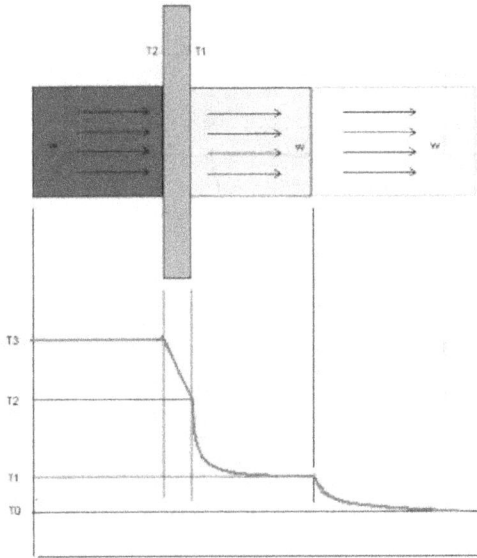

Figure-34
Internal winding and oil temperatures for the winding height are
shown in Figure-35

Fogure-35

A is the top oil temperature derived as the average of the tank outlet oil temperature and the tank oil pocket temperature.

B is the mixed oil temperature in the tank at the top of the winding (often assumed to be the same temperature as A).

C is the temperature of the average oil in the tank.

D is the average temperature at the bottom of the winding.

E represents the bottom of the tank.

G_r is the average oil to average winding temperature.

H is the hotspot factor.

P is the hotspot temperature.

Q is the average winding temperature determined by resistance measurement.

X axis indicates the temperature rise.

Y axis indicates relative positions.

measured point
Calculated point

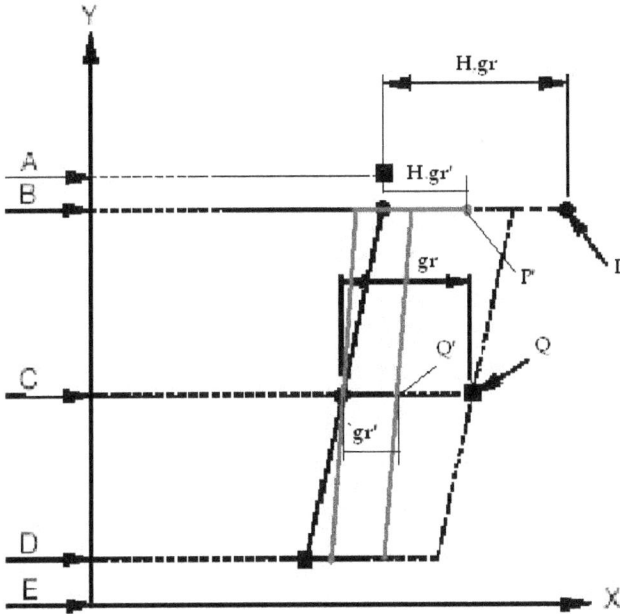

Figure-36

The effect of pumping oil is to reduce the winding gradient (gr); the mean oil rise stays the same but the drop across the cooler decreases, and the hotspot temperature rise (P) decreases. The effect is shown in Figure-36.

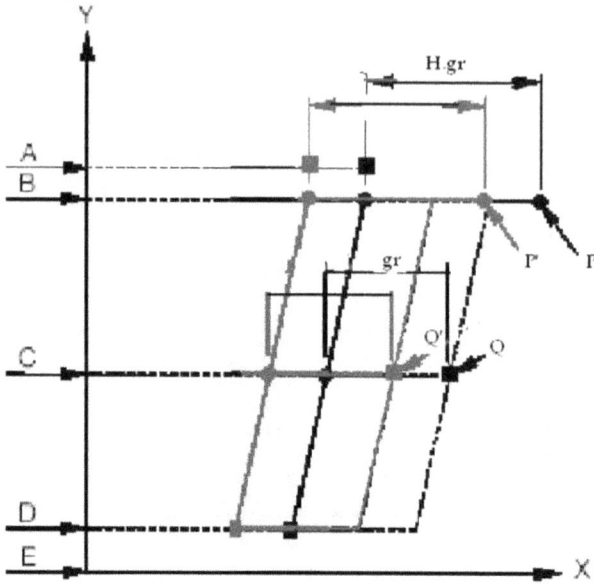

Figure-37

Figure-37 shows the effect of blowing air, this reduces the mean oil rise and the top oil rise, but the gradient stays the same.

Figure-38

Different cooler arrangements: Figure-38 is an example of OFAF cooling on GT

Figure-39
Figure-39 is an example of OFAN cooling on a large Auto

Figure-40

Figure-40 is an example of ONAN/ONAF/OFAD Cooling on a large Auto.

If the transformer has several windings then a thermal model considering all windings should be made. A thermal model is required for the core and winding in order to assess the hotspot temperature rise for each winding. The diagram assumes that the temperature of the oil entering the windings is at a single value for all windings and is shown as the bottom oil temperature. This common temperature will also apply to the core cooling oil;

providing that there is access to the core from the coolers. As the core and each winding have different quantities of heating, then the oil at the top of each of these will be at different temperatures. This oil will mix at the top of the tank on its route to the top of the coolers and the temperature known as the top oil temperature will be the weighted mean of all these temperatures.

The temperature gradient of the windings increases as a function of the current the winding, and this function is different depending on the type of oil flow. The IEC specifies the multiplying factor as $(I_1/I_2)^x$ where x = 1.6 for natural oil flow and x = 2.0 for forced oil flow.

The temperature rise of the oil increases as a function of the total loss in the transformer, and this function is different depending on the type of air flow. The IEC specifies the multiplying factor as $(W_1/W_2)^y$ where y = 0.8 for natural air flow and 1.0 for forced air flow. If the temperatures are known for one particular load, then they may be estimated for other loads and a set of cures may be drawn to estimate the performance for the various types of cooling. Typical hotspot values for various cooling and PU loading for a transformer assuming an ambient of 30°C. These are shown in Figure-41.

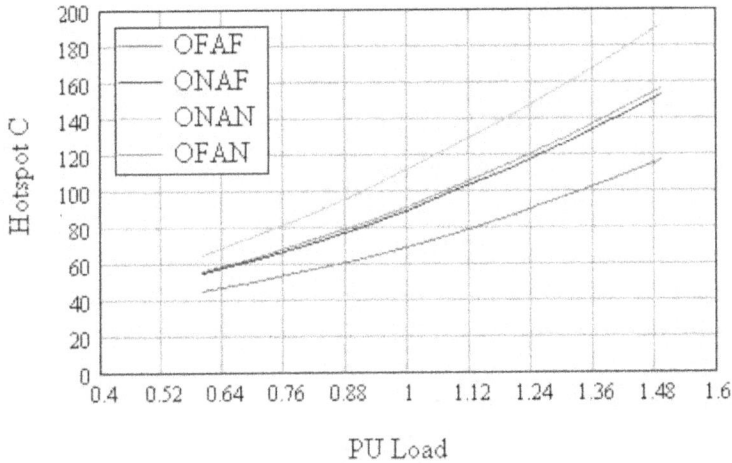

Figure-41

The design of a transformer is basically an iterative process with critical targets that must be met. The criteria are normally something that is calculated from the physical parameters of the design. The transformer as designed is viewed from different aspects in order to determine whether the design parameters are suitable to meet the required criteria. These aspects may be subdivided by considering the transformer as different models.

•Magnetic
•Mechanical
•Electric – Power Frequency
•Electric – High Frequency

Magnetic considerations

The Magnetic model considers the core and windings from a fully loaded point of view and the shape and magnitude of the leakage field is considered and calculated everywhere within the actual windings for each tap position. This leakage field within the windings in conjunction with the winding current subjects the conductor to mechanical forces which result from the basic relationship F = BI. It is therefore important to determine the magnitude and direction of the flux density (B) at all points in the winding and the current (I) at that point.

In general the most onerous current occurs during a fault condition where the current is only limited by the internal impedance of the transformer. Where $I_f = V_1/Z$. This is shown in Figure-42, where the only impedance restricting the fault current is the impedance of the transformer.

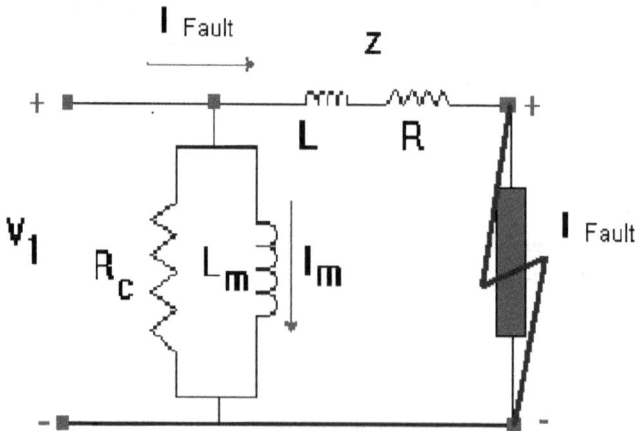

Figure-42

A typical leakage field plot is shown in Figure-43, near the top of the windings, for a two winding transformer. It can be seen that the field in the main gap is essentially axial and the field near the ends has an axial and a radial component. If a single conductor in the LV winding is considered, then the field line can be represented by two components, one axially and one radially, the radial field in the LV is directed towards the core and the associated force will be axial, the axial field will create an inward force on The LV conductor. As the HV current is in the opposite direction, balancing ampere turns, the radial force in the HV will be outwards.

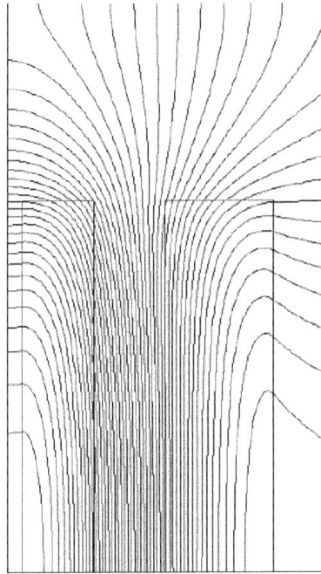

Figure-43

Consider a single flux line passing through the conductors, as shown in Figure-44.

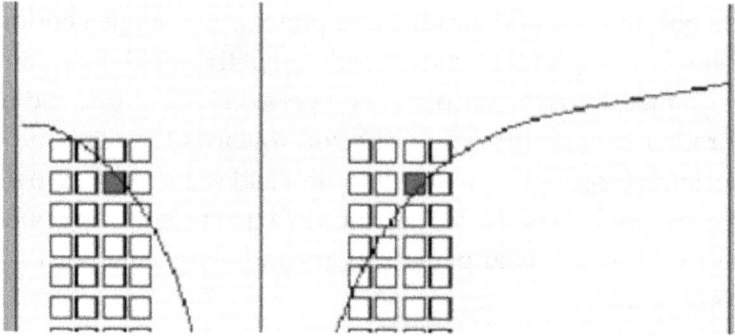

Figure-44

The current direction may be considered to be into the page for the LV and out of the page for the HV. Thus the direction of the forces generated is as shown. Note that the LV and HV currents are always in anti-phase and the ampere turns are balanced. To simplify the calculation the force, which is right angle to the flux and current, is shown as two components; radial and axial. The components are shown in Figure-45 for the conductors at the top and bottom of the LV and the HV.

Electro-magnetic Forces

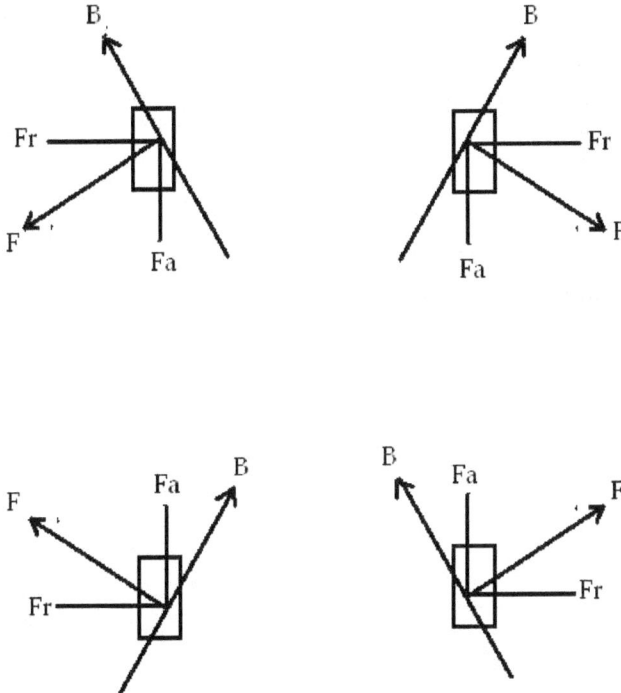

Figure-45

This will create a radial compressive force in the inner winding conductor and a radial tensile force in the outer winding conductor. The axial component of force, which is mainly at the ends of the winding.. This force will be accumulated throughout the winding height and will compress the axial insulation and may lead to tilting of the conductors.

Once the various forces are determined, by applying these to the conductor dimensions, the stresses in the conductor and associated insulation may be determined. These stresses are then compared to the strength of the material or critical stresses to ensure that the conductors can withstand these stresses without displacement.

Reference may be made to the IEC 60076-5, which will give guidelines to the calculation of the dynamic stresses and critical stress values.

Electric fields

The theory of electric fields is well documented in text books. This section has been written to discuss how the theory relates to High Voltage Power Transformers, and to give an insight into some of the issues and possible solutions to overcome the very high stress values that may exist in certain insulation structures.

Electrostatic calculation of Electric Fields

The fundamental unit used is the charge of an electron or proton which is;

1.601×10^{-19} Coulombs.

The types of substances are conductors, insulators and dielectrics. The volume resistivity (ρ) measured across unit volume, with the assumption that the current density is uniform is the basic difference between various substances and ρ varies between 10^{-8} ohm-meters for a conductor and 10^{18} for an insulator, but the assumption in electrostatic theory is that $\rho = 0$ for a conductor and $\rho = $ infinity for insulators and dielectrics.

Coulombs Law

The force between two charges at rest, acting along a straight line joining these charges, is proportional to the product of the charges and inversely proportional to the square of the distance between them.

Force = $Q_1Q_2/4\pi e_0 e_r r^2$ Newtons

Where:
Q_1 and Q_2 are in Coulombs
r is the distance between Q_1 and Q_2 in metres,
$e_0 = 8.85 \times 10^{-12}$ F/m,
e_r = relative permittivity

Electric Field

Is the region which surrounds electric charges and in which a small charged particle, introduced into the field experiences a force, the direction of the electric field is taken as that direction in which the particle would act if it were positively charged.

Assume that a small charged particle q enters a field E, then it will experience a force F where F=Eq Newtons.

Gauss' Theorem

The total outward normal flux from any closed surface is numerically equal to the sum of the enclosed charges.

$\Phi = \Theta$

The flux density (D) = Φ/a

$D = E\varepsilon_0\varepsilon_\rho$ and $\varepsilon_0\varepsilon_\rho = D/E$
For any given permittivity the ratio of electric flux density to the voltage gradient is constant.

Applications of Gauss' Theorem

Parallel Plates:

Consider two large parallel plates with a charge +q on one and –q on the other with a separation of d meters, and a permittivity of e, as shown in Figure-46. Then consider the charge density to be D, then E=D/e, and V = -Ed and the field is uniform. Each square is the same dimension and has the same flux and voltage.

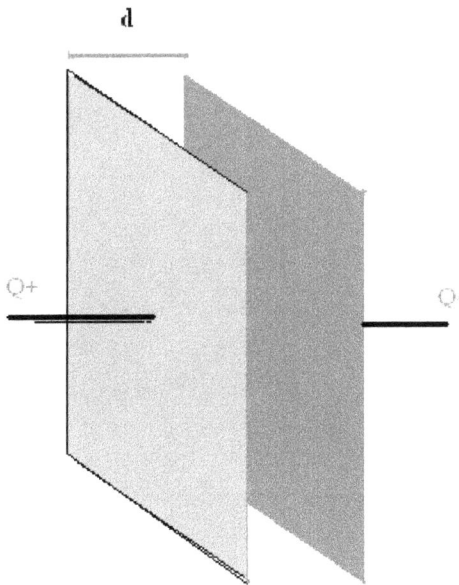

Figure-46

Concentric Cylinders

Consider two concentric cylinders of radii r and R, as shown in Figure-47. The charge on the inner cylinder is q coulombs per meter, at any radius 'x' the area is $2\pi x$.

$D = q/2\pi x$ and $E = q/2\pi x.e$ at position 'x'

Therefore both the density and the gradient are functions of 'x'

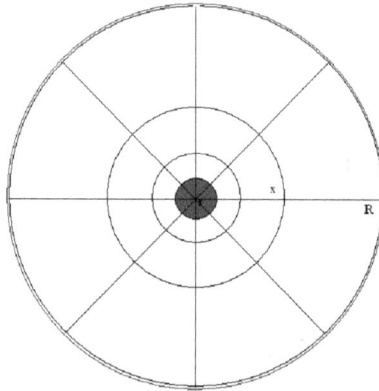

Figure-47

The highest flux density and the highest stress occurs when x = r, at the surface of the smaller radius. Note that each square has the same amount of flux and the same voltage and therefore the same amount of energy. These are sometimes known as curvy-linear squares, and this technique may be used to estimate by hand where the maximum stress occurs for any unusual electrode arrangement.

Parallel bars

Consider two parallel bars, and the electric flux lines at right angle to the equi-potentials, as shown in Figure-48.

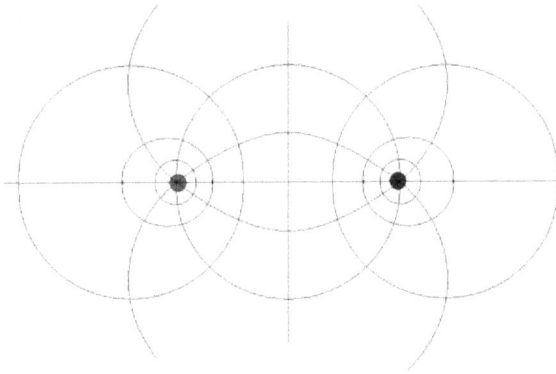

Figure-48

The density and the stress are highest at the small diameter, and each curvy-linear square has the same amount of flux and the same voltage difference.

Simple Field Plotting

By sketching the electrode arrangement and drawing the flux lines and equi-potentials at right angles, try to form curvy-linear squares, an approximate two dimensional field plot may be developed which will highlight the high stress areas. The general rule is that the flux lines always cross the equi-potentials at right angles. Further accurate assessments may be made using more sophisticated programs if required.

This method is very useful to assess the stress on lead exits and may be used to estimate the minimum radius required for HV leads, and where there are sharp corners.

Consider the arrangement shown in Figure-49, where the conductor boundaries form an internal corner, it can be seen that the highest stress is approximately twice the mean stress.

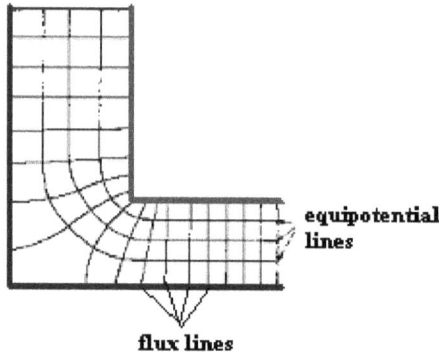

Figure-49

If the inner electrode represents the bottom of a winding at a position inside the core window, and the outer electrode represents the core and bottom yoke, the maximum stress will be from the inside corner of the winding. This stress may be relieved by rounding off the corner of the winding with the use of a stress ring, as shown in Figure-50.

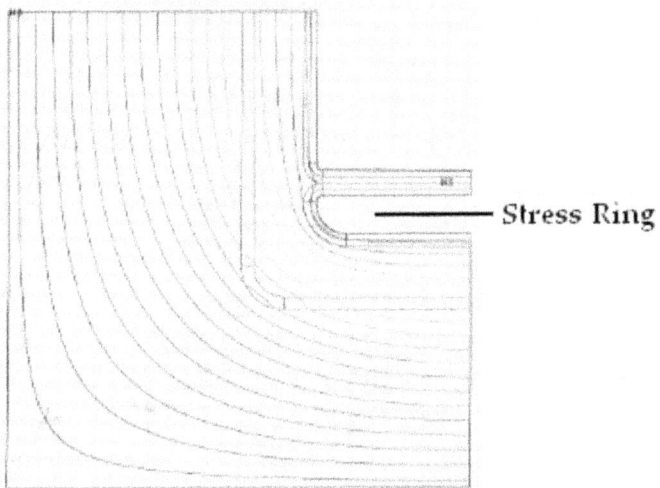

Figure-50

Insulation Strength

Once the electric field has been established and the maximum stress has been located, it is now the task to decide whether or not the insulation used can tolerate this stress, i.e. what is the strength in electrical terms.

The electrical strength of any insulation structure will depend on the shape of the voltage being applied, i.e. the curve of magnitude versus time, whether it is a short time, a long time, or a continuous repetitive wave such as a sin wave; the breakdown value of insulation is an inverse function of the amount of volt-seconds applied.

The general rule is that the strength is increased as the time is reduced. The strength for a continuously applied power frequency voltage is only about one third of the strength for a 1/50 microsecond Impulse Voltage. A switching surge wave of around 2000 microsecond only has 80 % of the strength for a 1/50 microsecond wave. These numbers vary for individual factories depending on the safety factors required and reference should be made to individual Design Standards to find the proper relationship.

Breakdown Mechanisms

The physics of breakdown is well documented and this section will only relate a few of the major differences in materials used in transformers. There are three main categories, breakdown in gases, solids and liquids.

Consider a composite dielectric with solid (pressboard), air (bubbles) and liquid (oil), formed as shown in Figure-51.

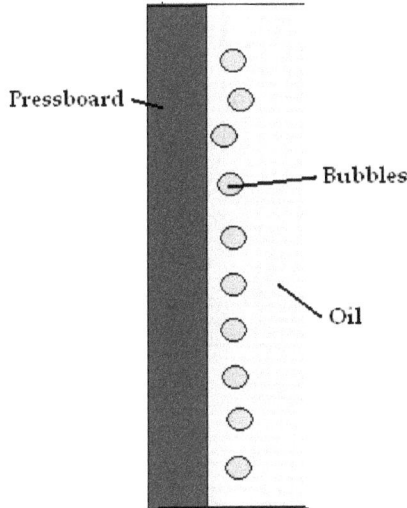

Figure-51

Each material will have a maximum allowable stress and for the materials used in transformers, gas breaks down in the region of 3 kV/mm, Oil will break down in the region of 30 kV/mm and solid will break down in the region of 60 kV/mm. The relative permittivities of these materials are approximately 1,2 and 4. If the materials are combined in a series composite insulation structure as shown in Figure-52, then the distributed voltage will follow the equation corresponding to the series capacitance of each part.

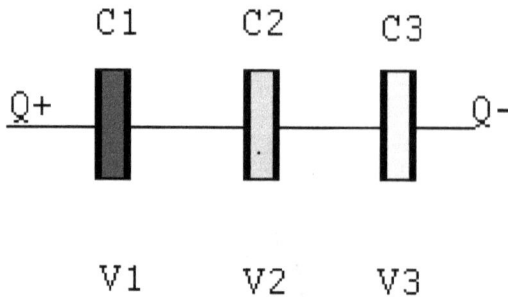

Figure-52

Consider that C1 represents the gas, C2 represents the oil and C3 represents the solid, then the capacitances assuming the sample cross section (a) is the same for each material.

$C1 = a\varepsilon_o\varepsilon_1/d_1$	$C2 = a\varepsilon_o\varepsilon_2/d_2$	$C3 = a\varepsilon_o\varepsilon_3/d_3$
$V1 = Q/C1$	$V2 = Q/C2$	$V3 = Q/C3$
$E1 = V1/d1$	$E2 = V2/d2$	$E3 = V3/d3$
$E1 = Q/a\varepsilon_o\varepsilon_1$	$E2 = Q/a\varepsilon_o\varepsilon_2$	$E3 = Q/a\varepsilon_o\varepsilon_o$
$\varepsilon_1 = 1,$	$\varepsilon_2 = 2$	$\varepsilon_3 = 4$

Then the stress in the gas is 4 times the stress in the insulation and 2 times the stress in the oil, and since the electric strength in the gas is one tenth of the strength of the oil then the gas will breakdown forty times easier than the oil. That is because bubbles are very weak electrically and must be avoided. Even if the bubbles are very small, whilst they may not cause a problem with the surrounding oil or solid, they will still breakdown and the associated arcs will most certainly display electrical discharge, or Pd.

A well designed oil-filled transformer theoretically should not have any gases, however there are certain areas which, during a short overload can overheat and it is known that transformer oil at 130 °C can generated bubbles; these bubbles are gaseous and have a relative permittivity of 1. If this area is subjected to an electrical stress, then a weak point will be created and possible breakdown can occur in the gaseous region. The bubbles may be generated in an area where there is no major stress such as earthed clamps, however the bubbles can migrate into the winding region where a high stress does exist, and discharge or breakdown can result.

Consider the effect of moisture in a composite insulation structure as shown. This time the relative permittivity of the water is very high, around 80 and it may be treated as a conductor, allowing an easy path for discharge particularly in the axial direction along the surface of cylinders or wraps. See Figure-53.

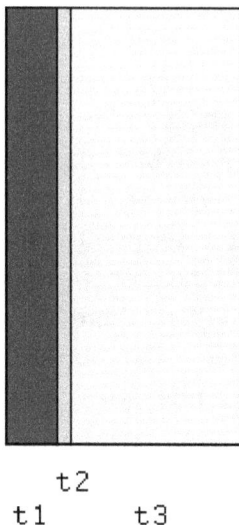

t2
t1   t3

Figure-53

If there is an axial stress along this water path, then discharge will occur and the surface will start to carbonize leading to a breakdown.

Sometimes in multi-duct insulation the cylinder which forms an oil duct can be dried by circulating oil, but the inner wraps may not be in the flowing oil path as shown in Figure-54, especially for OD type cooling, and care must be taken to dry the complete insulation structure prior to any high voltages being applied. Once the onset of discharge due to moisture occurs, the insulation will have permanent tracking marks and it is not possible to recover from this situation.

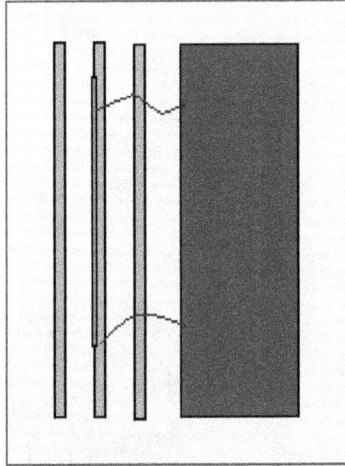

Figure-54

If this weakened insulation due to moisture is associated with a change in structure, such as a scarfed joint it can lead to a catastrophic breakdown as shown in Figure-55. The main glue line may also be subjected to changes in dielectric material, thus creating local high stress areas. It is important to note the electric strength of an insulation system is greatly reduced if there is a creepage path along the surface of the insulation. The strength is approximately half of the solid puncture strength. This must be taken into account when designing support boards for leads.

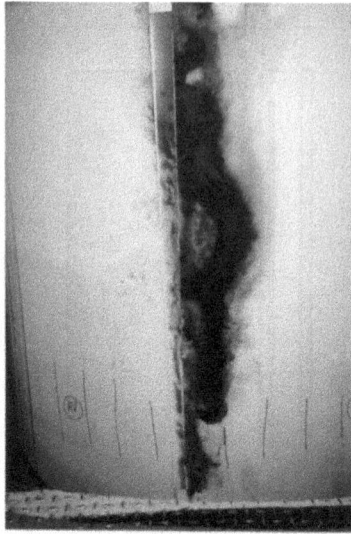

Figure-55

The size of the oil gap will have an effect the breakdown strength of the oil, and it has been observed that small oil gaps have a much higher strength than large oil gaps. It is normal therefore to sub divide any large gaps into a series of smaller gaps by introducing pressboard wraps. The wraps will of course take up some of the space and effectively reduce the oil gap, as the wraps have a higher permittivity the stress within the wraps will be smaller than the oil stress, it is therefore important not to use too much insulation as it may defeat the purpose. The effective oil gap may be determined by considering the total gap, the total solid and the remaining oil gap and adjusting for the increased stress in the oil. Consider Figure-56.

100 kV

100 mm **30 oil   70 solid**

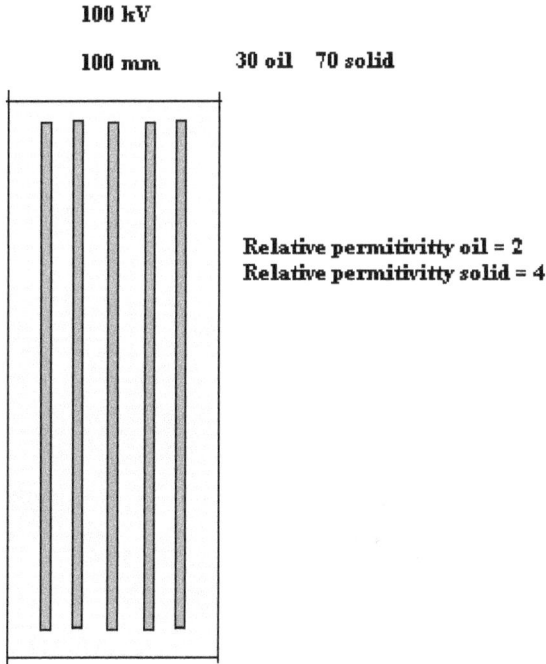

Relative permitivitty oil = 2
Relative permitivitty solid = 4

Figure-56

Example;

Main gap =100 mm
Solid component = 30 mm permittivity = 4
Oil component = 70 mm permittivity = 2

Voltage across main gap = 100
Q=CV = C1V1 = C2V2
1/C = 1/C1 + 1/C2
V = V1 + V2

C1=4/30 = 0.1333
C2=2/70 = 0.0286

1/C =30/4 + 70/2

1/C = 7.5 + 35

C = 0.0235
Q = 0.0235 * 100 = 2.35

V1 = 2.35/0.1333 = 17.6
V2 = 2.35/0.0286 = 82.4

Stress in oil = 82.4/70 = 1.178
Stress in solid = 17.6/30 = 0.587
Average stress = 100/100 = 1.0

Therefore by introducing 30 % solid insulation into the gap the oil stress has increased by 17.8 %. It is therefore important that the oil strength due to the smaller ducts increases by more than 17.8 %, otherwise there is no advantage.

For estimating purposes the equivalent oil gap has reduced from 100 mm without any insulation to 100/1.178 = 85 mm.

Breakdown strength of small oil ducts.

In general the intrinsic strength of dry, de-gassed oil is very high about 50 kV/mm for very small ducts, it is however not practical to use these values in transformers and it is more realistic to use oil gaps in the regions of a few mm to a few hundred mm. The oil strength across this range varies from about 12 kV/mm to 3 kV/mm, and each individual manufacturer will have preferred limits depending on the manufacturing dimensional control exercised. As previously stated the voltage wave shape will also have an effect on this strength, for estimates the impulse strength is about three times the power frequency strength.

The effect of the volume of oil under stress is shown in Figure-57. This is approximate for good quality oil showing a 50% chance of failure.

Oil Volume effect

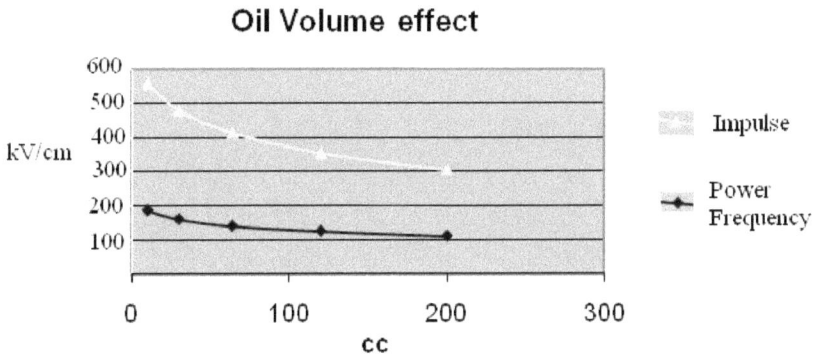

Figure-57

Reducing the volume of oil under stress is a common procedure and not only subdividing main winding gaps by using wraps, but this technique may also be used to improve the allowable strength around a small diameter tapping lead, it may be less expensive to use spaced wrap around a lead, rather than increase the diameter of the copper lead. See Figure-58.

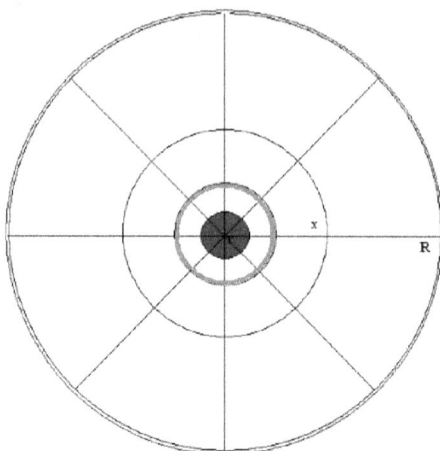

Figure-58

Insulation systems

There are normally different types of insulation associated with transformers and they may be classified as follows;

Major winding insulation – this is the insulation that is used to separate the individual windings, both between phases and in the main winding ducts including barriers to the tank.

Minor winding insulation – this is the insulation within a winding including turn-turn, section – section, group-group and layer – layer.

Lead insulation – this is the insulation between the leads to the tank, lead-lead and any turret barriers required.

The amount of insulation required is determined by the allowable voltage stresses that the insulation can safely withstand, the critical values usually have two or more levels as discussed in the previous section, and they are dependent on the duration of the voltage.

Power frequency voltages – usually up to 300 Hz, covering normal 50-60Hz operation, switching oscillations during normal CB operation which can induce twice normal volts per turn, induced and applied test voltages.

There are various power frequency test levels at different time durations. It may be pertinent to review the history of these test levels. In the beginning, the power frequency test was a simple, one minute test at twice the normal frequency (power frequency); this test was conducted by increasing the voltage slowly to the test level, keeping it there for 6000 cycles (for a 50 Hz supply) and reducing it slowly to zero. The criterion was that the unit did not fail. As the test systems became more sophisticated, techniques were developed to measure partial discharge. Pd measurement were then carried out during the rise at about 100% normal voltage rating, and repeated during the fall at 100% voltage rating. This gave an indication of any hysteresis effect created by the test. At that time there was usually a lot of external corona at the full over-potential level.

As these techniques developed, partial discharge became more important and in order to make proper measurements, longer duration tests were called for to allow any discharge to develop. Eventually the one-minute test was omitted and the preference for longer duration tests with Pd measurements was introduced. Hence the present 30-minute test at a reduced voltage with a 5 second period at the higher test level.

It should be noted that many companies developed critical insulation test levels in their laboratories using the 1-minute test levels, and today many still continue with this practice. One advantage was that the higher over-potential test level covered the twice-normal volts per turn that can occur during normal switching operation. It should be noted that conversion factors between one minute and thirty minutes are sometimes taken as unity. This should be avoided as the criteria for passing test is different, in the one case the unit should not fail, in the other case the unit should not discharge (or the discharges should be well within the specification).

The IEC specifies two levels of allowable Pd for HV units, 300 pC at 1.3 Um/√3 and 500 pC at 1.5 Um/√3.

The wave-shapes to define transient voltages – such as impulse and switching surge, chop and front of wave (FOW) tests are defined in the IEC specifications. The definition of the basic impulse wave is shown in Figure 59.

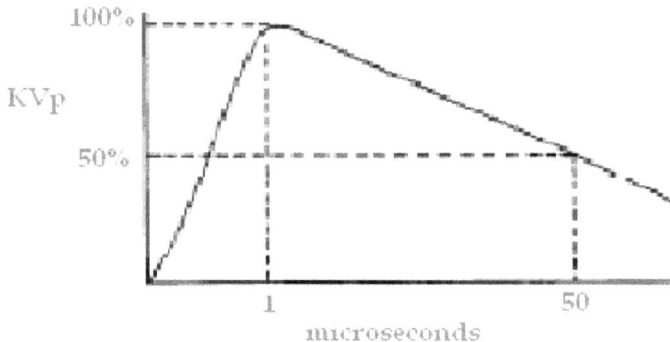

Figure-59

It is advisable to create a table, as shown in Figure-63, showing the critical voltages defined in Figure-60; one column representing the most onerous equivalent one-minute power frequency voltage, and one column showing the equivalent 1/50 microsecond impulse voltage. The winding insulation structure is then configured so that all the stresses determined by the final voltages and clearances are within the acceptable limits. The voltages shown in the winding clearance diagram will have to be calculated for the two types of voltage waveforms shown. Thus voltages V1 to V13 inclusive should be determined for both conditions.

The equivalent 1/50 voltage is usually derived by considering conversion factors for other transient type values, such as waves with shorter duration, chop waves – with a 3-5 microsecond duration, and front of wave chops- with less than 1 microsecond duration, will be less onerous than a full 50 microsecond wave. Approximate conversion factors are 1.1 and 1.15 respectively. That is a chop wave at 10% above the full wave value will require the same insulation strength as the full wave. Waves with a longer duration such as switching surge waves at 2000 microseconds will be more onerous than the 50 microsecond full impulse value and the equivalent conversion factor will be 0.8. That is the insulation required for a full wave is only 80 % of the insulation required for a switching surge wave. It is therefore necessary to apply all these conversion factors to the various test durations to determine which the most onerous voltage is, select the critical values, or safe working values from your design standards, and design for these.

Note that the critical values will depend on the electrode shape, the insulation material, whether you use local stress or average stress, allowance should be made for creep conditions, and the insulation is processed properly. As a guide the average radial stress between concentric cylinders is about 150 kVp/cm at the middle and 130 kVp/cm at the ends. The average 1 min power frequency stress in the radial dimension is 50 kV RMS /cm. Where there are sharp corners such as a core packet, this should be reduced depending on the electrode shape, or an electrostatic shield fitted. The critical stress for power frequency should also be reduced if very low partial discharge values are specified.

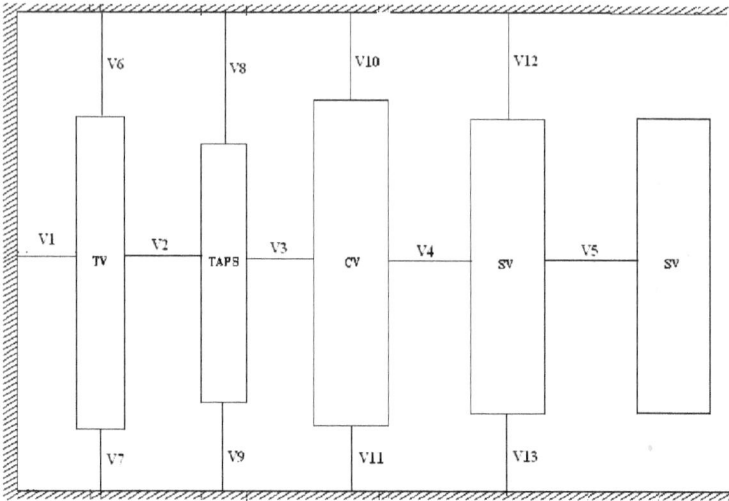

Figure-60

Due to the non-linear nature of the transient type voltages it is always pertinent to determine the axial and radial creepage stresses in the major insulation and compare these to acceptable limits.

The strength of oil is somewhat dependent on the volume under stress. Typical strength values for bare electrodes, both for Impulse and power frequency voltages are shown in Figure-61.

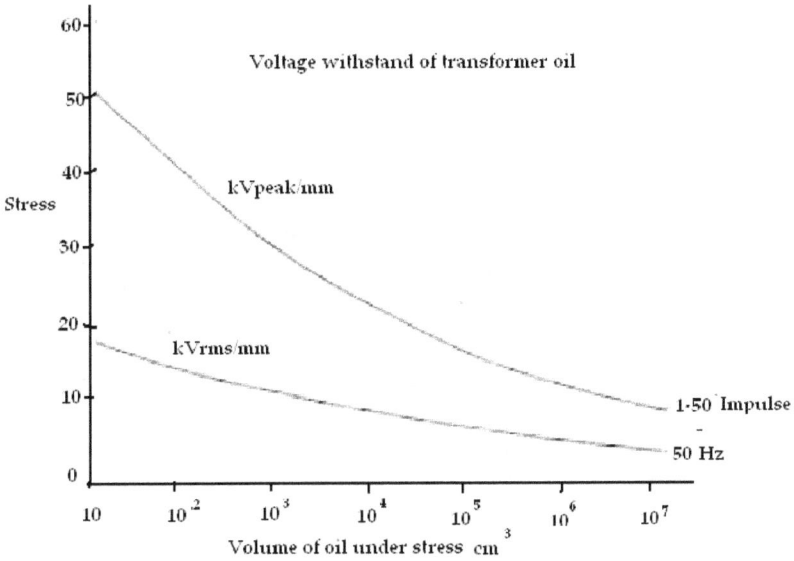

Figure 61

Surface creep strength of processed pressboard, is a function of the area under stress and typical strength curves are shown in Figure-62.

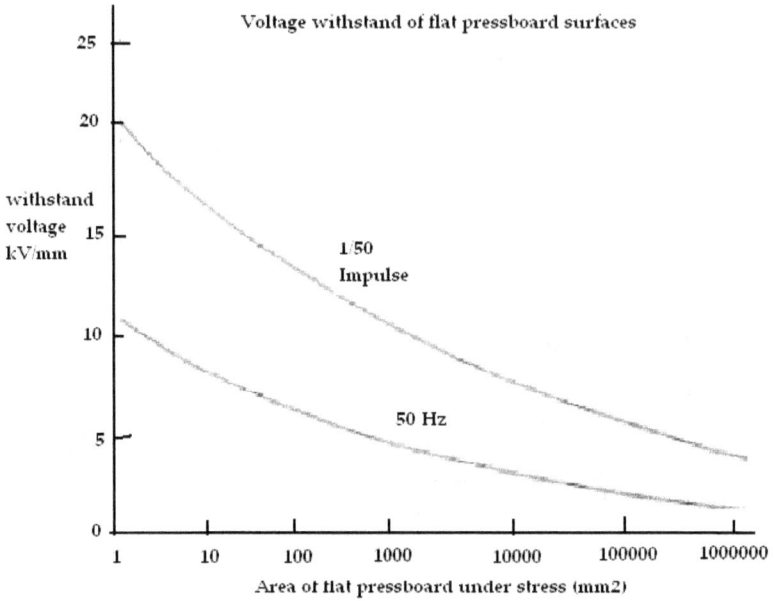

Figure-62

Voltage	Power Frequency	Equivalent 1-min	Impulse	Switching Surge	Equiv 1/50
V1					
V2					
V3					
V4					
V5					
V6					
V7					
V8					
V9					
V10					
V11					
V12					
V13					

Figure-63

A typical example of major insulation for the above winding arrangement is shown below in Figure-64, note how the main ducts and large oil spaces are broken up into smaller ducts, as this allows for a higher critical stress to be used. Note also the use of stress rings in the regions where there are high gradients; this feature reduces the local stresses at the winding ends by increasing the effective radius of the corners.

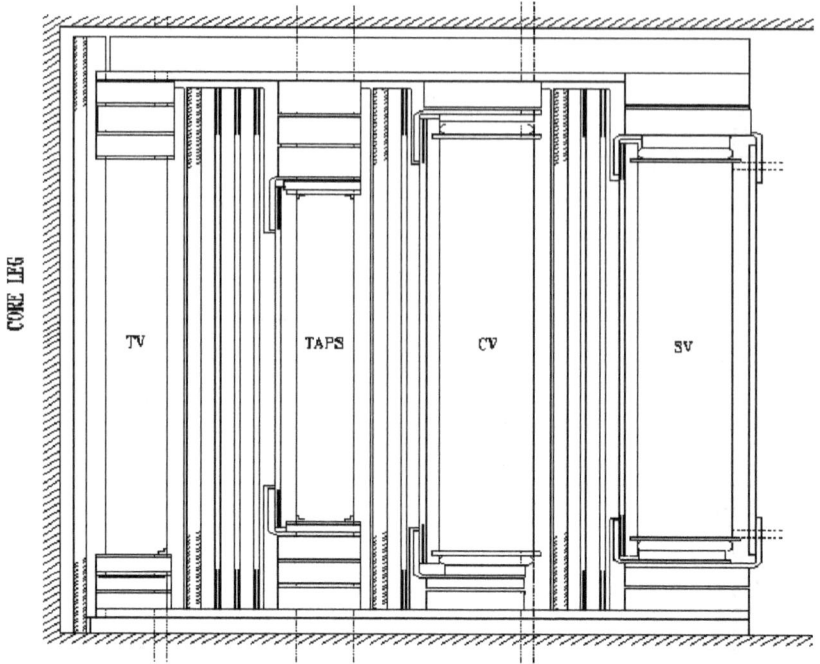

Figure-64

The major insulation structure must also be suitable to withstand the mechanical stresses that occur when the unit is processed or lifted or subjected to short circuit forces, and must therefore be very robust. The structure must also be able to allow the flow of cooling oil through the structure, hence the top and bottom block and washer arrangements.

The calculation of the series and shunt capacitances will depend on the disposition of the turns within the winding and the gaps between the windings and to the core and tank.

Series capacitance between turns.

The capacitance per turn between two adjacent conductors may be estimated by considering the conductor dimensions and material permittivity. Consider Figure-65.

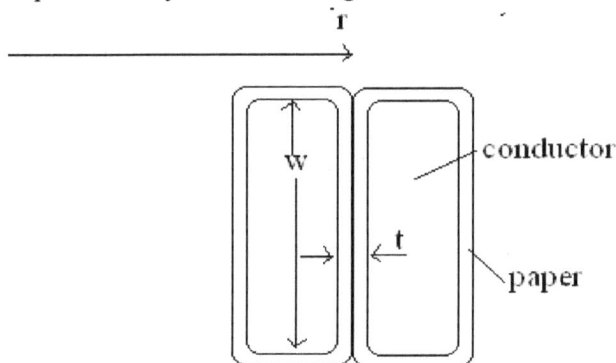

Figure-65

$$Ct = 2\pi r(w + t).\varepsilon_o\varepsilon_r / t$$

Open Disc Windings

The continuous disc winding is very popular for large power transformer windings: it offers the designer options for radial cooling ducts and high series capacitance. As the winding is continuous, without any breaks or connections, is can employ continuously transposed cable. The arrangement of the turns for one group is shown in the Figure-66, and as there is a facility to introduce transpositions, parallel conductors do not present a problem. It is more convenient to wind this type of winding on a vertical lathe as this eases the disc support during manufacture.

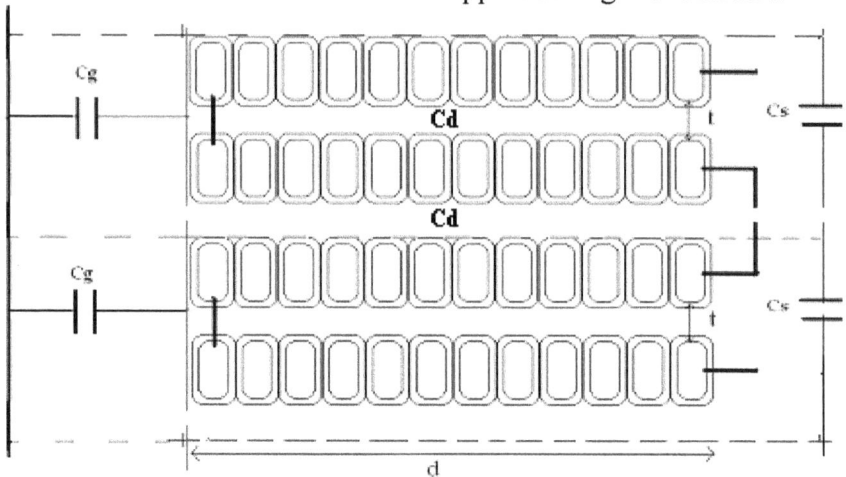

Figure-66

The series capacitance of this group of turns will include a contribution from the equivalent capacitance between the sections groups and turns.

R is the mean radius of the section

t is the distance between the conductors, which may be oil or solid or a combination of both.

It is initially assumed that the discs are not connected and the capacitance between the discs is a function of the dimensions and the dielectric constant given by;

$$Cd = \frac{2\pi r \times d \times \varepsilon_o \varepsilon_r}{t}$$

Due to the connection between the discs, the effective contribution to the series capacitance Cs will be Cd/3.
The series capacitance may be determined as follows;
The inter turn capacitance is Ct and the voltage across the paper is V (volts per turn).
The energy stored between the turns is

$$\xi t = (N - 1) \times \frac{Ct}{2} \times V^2$$

The energy stored between each pair of discs is

$$\xi d = \frac{Cd}{6} \times (NV)^2$$

If the series capacitance is Cs and the voltage across this capacitance is NV, then the energy stored is

$$\xi s = \frac{Cs}{2} \times (NV)^2$$

equating the energy stored

$$\xi s = \xi t + 2 \times \xi d$$

From which the series capacitance for the group is Cs where;

$$Cs = 2 \times \frac{C}{3} + Ct \times \frac{(N-1)}{N^2}$$

The shunt capacitance for the group is given by the total shunt capacitance for the winding divided by the number of groups.
The insulation required between the turns is paper and the thickness of the paper is controlled by V. The insulation between sections in the group is controlled by N times V and the initial distribution when impulse voltages are applied to the winding. The insulation between the sections may consist of a composite arrangement of paper, oil and pressboard.
This initial distribution should also be considered when designing the major insulation as local stresses may create creep paths in insulation close to the winding such as oil cooling ducts.

In order to determine the local axial stress in this region it is necessary to determine the initial voltage across the first few sections and assume that this voltage is distributed along the physical dimensions of the first few sections, rather the actual gap between the sections.

Tap Windings

Spiral windings are also used where an internal tap winding with several sections is required. In this case the winding sections are all wound in parallel using a multi-start arrangement; they are then connected in series using external leads, often known as 'organ pipes' due to the similarity of construction. An example of the interconnections is shown in Figure-67.

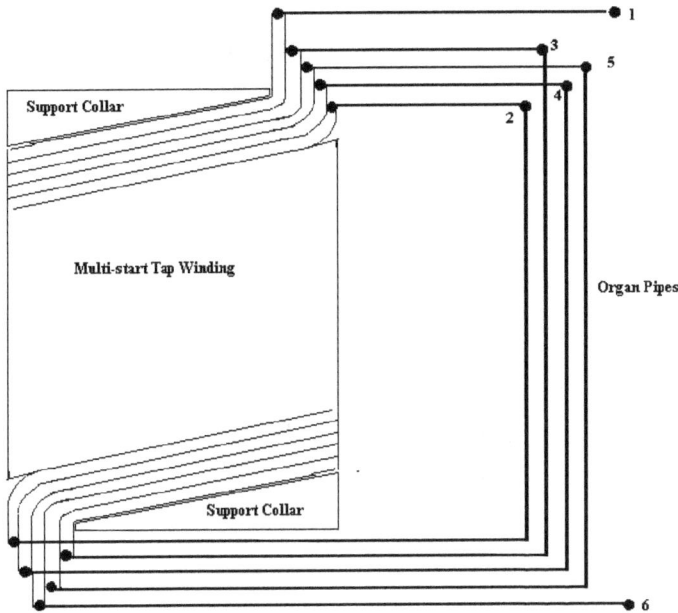

Figure-67

Interconnecting in this fashion will restrict the operating voltage across the conductor insulation to no more than twice the section voltage. This arrangement will also have a high series capacitance as the voltage between adjacent conductors is that of two sections, and this will help the initial capacitive distribution during impulses.

Where;

Capacitance per turn between adjacent conductors

$$Ct = 2\pi r(w + t).\varepsilon_o \varepsilon_r / t$$

For V volts per turn, the energy stored per turn

$$\xi t = \frac{Ct}{2} \times (2NV)^2$$

The energy stored per section with N turns and 2NV volts across the paper is

$$\xi s = N \times \frac{Ct}{2} \times (2NV)^2$$

The energy stored per section with NV volts and Cs as the equivalent series capacitance.

$$\xi s = \frac{Cs}{2} \times (NV)^2$$

Equating the energies give an equivalent series capacitance per section

$$Cs = Ct \times 4N$$

Disc windings can also be interleaved as shown in Figure-68.

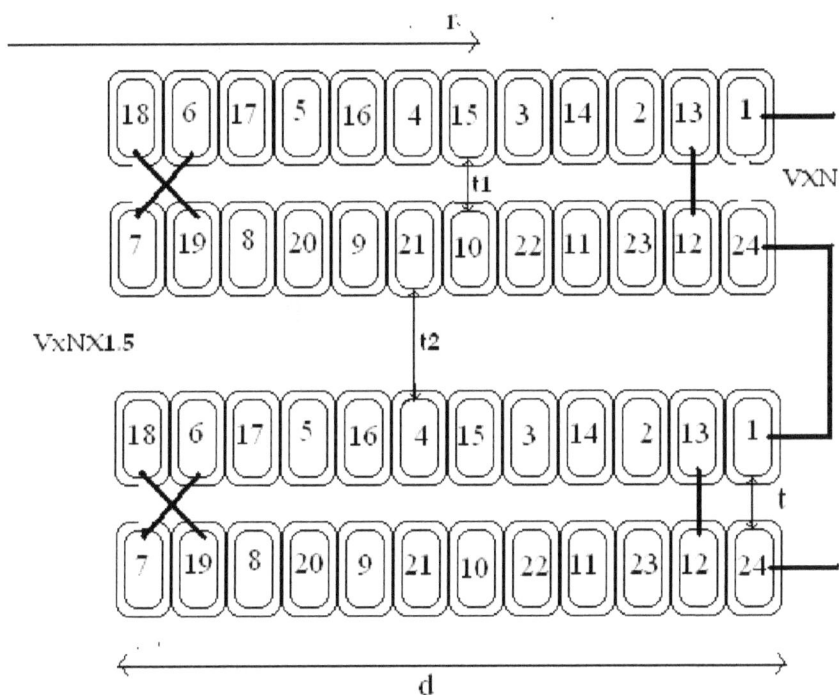

Figure-68

The capacitance between adjacent conductors is the same as for a disc winding

$$Ct = 2\pi r(w + t) \cdot \varepsilon_o \varepsilon_r / t$$

In this case the voltage across the paper is VN/2 where N is the turns per group, and the energy stored in the group turns will be

$$\xi t = \frac{Ct}{2} \times N \times (\frac{VN}{2})^2$$

There will also be some energy stored in the spaces between the sections as per the continuous disc.

$$Cd = 2\pi r \times d . \varepsilon_o \varepsilon_r / t$$

If t1 and t2 are different with different permittivity then

$$Cd = 2\pi r \times d \times \varepsilon_0 \times (\varepsilon_{r1} / t_1 + \varepsilon_{r2} / t_2)$$

The energy stored between these sections

$$\xi d = \frac{Cd}{2} \times (\frac{NV}{2})^2$$

The sum of the energy store for the group

$$\xi s = \xi t + 2\xi d$$

And

$$\xi s = \frac{Cs}{2}(NV)^2$$

equated to an effective capacitance give an effective capacitance with an applied group voltage NV is given by;

$$Cs = Ct \times \frac{N}{4} + \frac{Cd}{2}$$

If the turns in the two sections are interleaved as shown, then the voltage between the adjacent conductors is NV/2. This improves the overall series capacitance significantly and hence reduces the α factor in the calculation for the initial impulse distribution.

The disadvantage being that there will be more conductor paper required and the voltage between the groups will be 1.5 times the voltage between the sections.

Due to the complexity of the interleaved group, it is common to restrict this type of winding to a few groups at the line end, and use continuous disc for the remaining groups.

This initial distribution should also be considered when designing the major insulation as local stresses may create creep paths in insulation close to the winding such as oil cooling ducts. In order to determine the local axial stress in this region it is necessary to determine the initial voltage across the first few sections and assume that this voltage is distributed along the physical dimensions of the first few sections, rather the actual gap between the sections.

As a further explanation consider a disc winding as shown in Figure-69, the voltage across the first group is 100 kVp, the voltage across the second group is 85 kVp. This calculation is based on a mathematical model, which uses a calculated series capacitance per group, and an effective ground capacitance per group. The plates of this fictitious capacitance lie at the centre line of the two inter-group spaces and the voltage gradient across this capacitance is distributed between these two plates. It is this voltage and this distance that determines the maximum average axial stress in the adjacent insulation, which is treated as a creep stress.

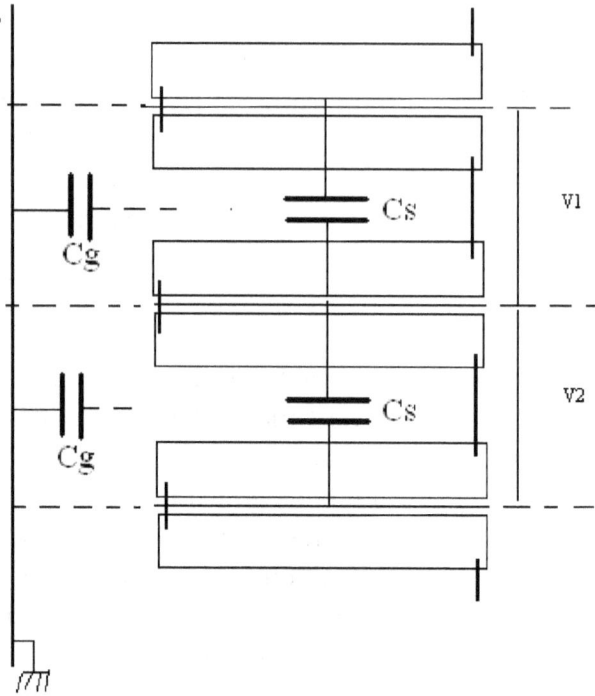

Figure-69

The two capacitances Cs and Cg are estimated from the physical dimensions of the winding, and consist of the sum of various capacitances within the boundaries chosen. For example, the Cs as shown above will be determined from the capacitance between the section and the capacitance between the groups and the capacitance between each turn. As each has a different voltage the effective capacitance is determined by adding up all the stored energy levels and equating them to the effective voltage and capacitance per bit. The capacitance Cg will consist of the inner and outer capacitance per bit. The inner capacitance will be to the core or an inner winding and the outer capacitance to the tank or to an adjacent outer winding.

The same technique of equating stored energy is used to determine the initial impulse distribution when the winding is a simple spiral, a multi-start spiral or an interleaved winding.

Note that the same dimensions can be used when considering the power frequency creep stress and this should not exceed the agreed value.

In most cases the voltage V1 is taken as the voltage existing between the first two sections at the front and the first two groups at the back. This will determine the section-section and group-group insulation within the winding. Note that for an interleaved winding the voltage between the groups is one and a half times the voltage between the sections.

Impulse distribution, calculation for partial interleaving of windings.

The Impulse wave can be considered in two sections, the first section is the rise time of the wave which essentially has very high frequency components and the distribution of the winding is dictated by the capacitance effect. The inductive reactance of the winding at these frequencies is deemed to be very high and may be neglected.

For the initial distribution the winding can be represented as a chain of capacitors in series (Cs) with a capacitance to earth (Cg) at each node, as shown in Figure-70.

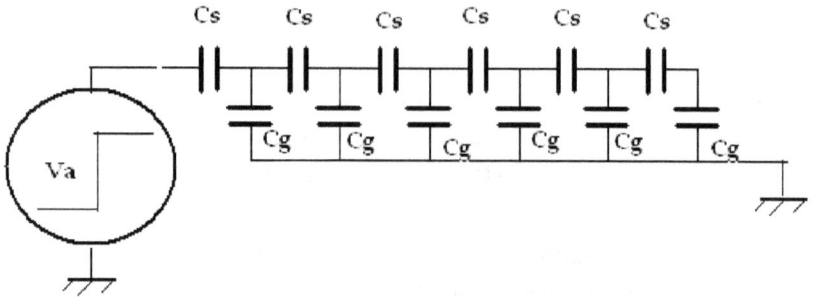

Figure-70

The applied voltage (Va) can be represented by a square wave pulse, and the distribution of the voltage at each node is non-linear. The effect of this non-linearity can be represented by a function of the ratio between the series and ground capacitances. This function is termed the α where;

$$\alpha = \sqrt{\frac{Cg}{Cs}}$$

If the winding has N groups then the α for the winding is given by;

$$\alpha = N \times \sqrt{\frac{Cg}{Cs}}$$

The distribution of the voltage along the winding for a uniform winding with the applied voltage at one end and the other end grounded is shown in graphical form for various values of α.

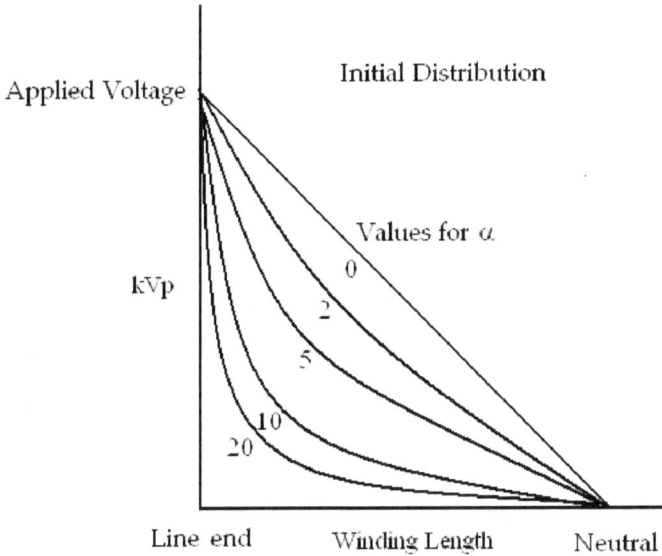

Figure-71

As can be seen from this set of curves, as the α factor approaches zero, the distribution is linear and follows the turn's distribution along the winding. In this case the Cg is very low and the Cs is very high.

It is clear from the curves that the voltages near the line end sections are more onerous than the voltage gradients in the middle of the winding, and reinforced insulation or a wider spacing between the discs will be required at the line end sections.

The initial distribution along a winding with 'N' sections and Impulse level E, is given by considering the voltage at section x as;

$$Vx = E.\frac{\sinh\left(\dfrac{\alpha(N-x)}{\alpha}\right)}{\sinh(\alpha)}$$

As time passes the initial distribution changes, the high frequencies are replaced with the lower frequencies and the main distribution becomes inductive and proportional to the turns. There is a period in between where there may be some oscillations and resonant frequencies can create voltages that are higher than the initial voltages.

The envelope for this oscillation may be estimated by adding the difference between the steady state voltage at section x and the initial voltage at section x, to the steady state voltage at section x, as shown in the sketch Figure-72.

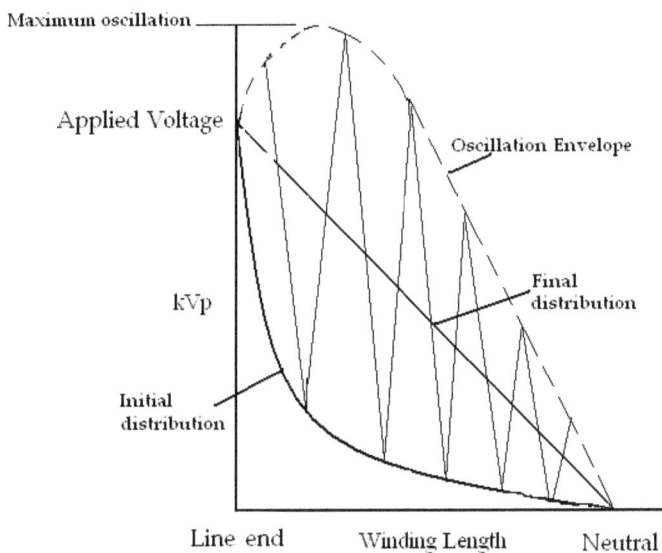

Figure-72

A winding with a separate tap situated at the line end, may have an oscillation that will require earth clearance higher than the test voltage. See Figure-73.

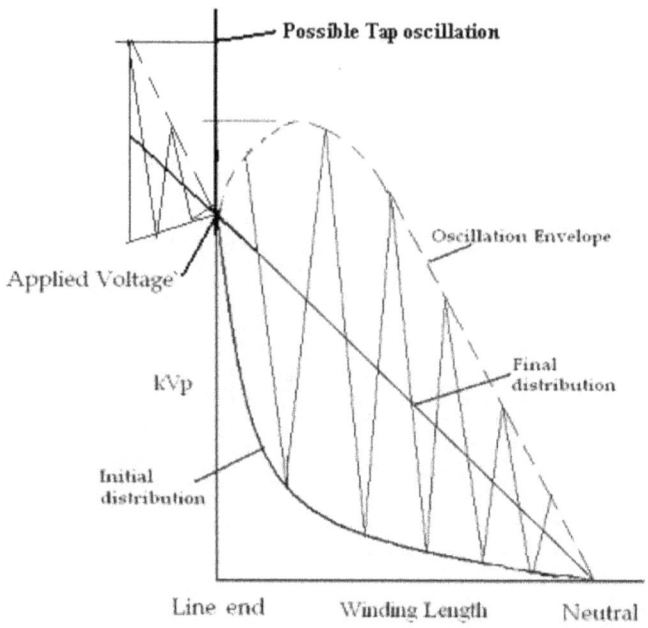

Figure-73

If there is a tap winding at the neutral end, and then if it is overhanging, there may be an oscillation as shown in Fugure-74, and suitable insulation should be provided.

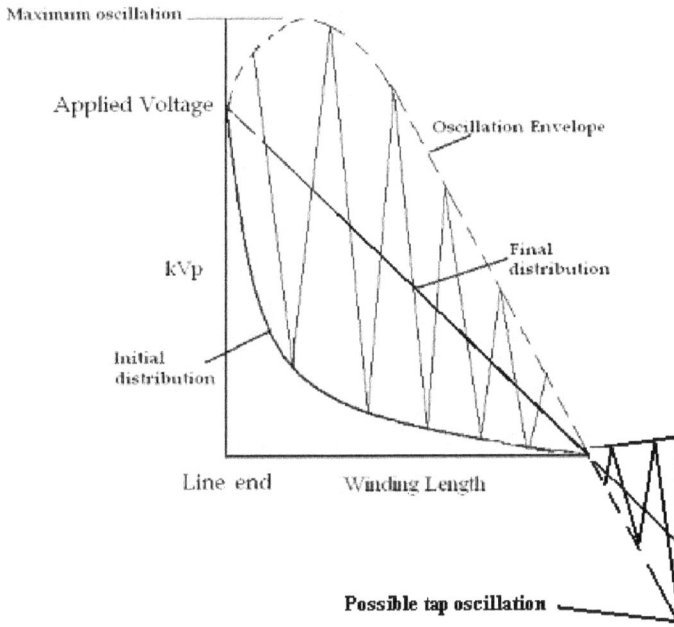

Figure-74

In the event that the winding has some grading, for example interleaved at the line end and continuous disc for the neutral end, there will be two values of α. The initial distribution will therefore have two parts and the oscillations may have two envelopes as shown in Figure-75.

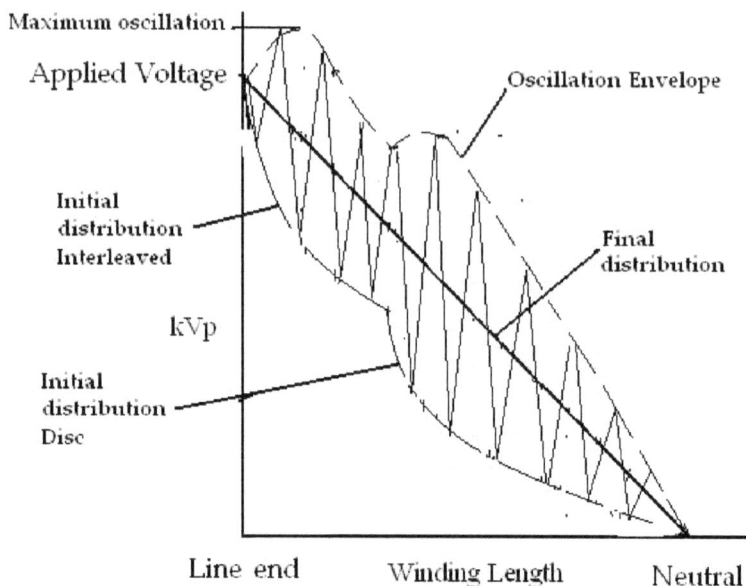

Figure-75

The calculation for this initial distribution requires that the input capacitance for the continuous disc section is calculated and used as a terminating capacitance for the interleaved section. Consider the winding to be made in two parts, the first part nearest to the neutral is part 1 and the second part nearest the line end is part two. There will be three nodes, node 1 will be the neutral, node 2 will be the junction and node 3 will be the line end.

The input capacitance $Cj2$ at the junction of a winding terminated by a capacitor $Cj1$ is given by;

$$Cj2 = \frac{Ck1 \times [\sinh \alpha1 - Cj1 / Ck1 \times \cosh \alpha1]}{[\cosh \alpha1 - Cj1 / Ck1 \times \sinh \alpha1]}$$

Where

$$Ck1 = \sqrt{Cs1 \times Cg1}$$

if the neutral is directly earthed Cj1 will be a very high number, and the junction capacitance will reduce to

$$Cj2 = Ck1 \times [\cosh\alpha1]/[\sinh\alpha1]$$

The input capacitance at node3 may be determined by

$$Cj3 = \frac{Ck2 \times [\sinh \alpha2 + Cj2 / Ck2 \times \cosh \alpha2]}{[\cosh \alpha2 + Cj2 / Ck2 \times \sinh \alpha2]}$$

Where

$$Ck2 = \sqrt{Cs2 \times Cg2}$$

The initial voltage at the line end will be E, or V3 and the initial voltage at the junction V2 may be determined using the ratio of the winding input capacitances

$$V2 = V3 \times \frac{Cj2}{Cj3 + Cj2}$$

Consider the voltage at section x in the first part with N sections;

$$Vx = V2. \frac{\sinh\left(\frac{\alpha1(N-x)}{\alpha1}\right)}{\sinh(\alpha1)}$$

Consider the voltage at section x in the second part with N sections

$$Vx = (V3 - V2) . \frac{\sinh\left(\dfrac{\alpha 2(N - x)}{\alpha 2}\right)}{\sinh(\alpha 2)} + V2$$

The composite distribution will then follow the pattern shown in Figure-75.

The same technique can be used to determine the voltage distribution for multiple changes within the same winding. These calculations may be used in a small program using BASIC or MATHCAD which can handle multiple subscripts and make the calculations very quickly. They are useful to compare various iterations and combinations of part interleaving.

There are more complex programs that can handle input for multi-terminal arrangements which can show how the initial distribution is transferred to other concentric windings, but these are mainly used when the design is complete for verification purposes.

Transient voltages

As discussed a little in section 6, transient voltages occur in an HV system when a circuit breaker is closed and depending on the precise time of application and the remanent flux in the transformer core, over-flux conditions will be set up end higher than normal voltages will exist in the transformer windings until the core flux stabilises back to normal.

There are also re-strikes that occur during CB opening whilst the contacts are in motion, the gap is increasing and becoming more tolerant to the applied voltage but before the contacts are fully open, re-striking will occur as shown in Figure-76.

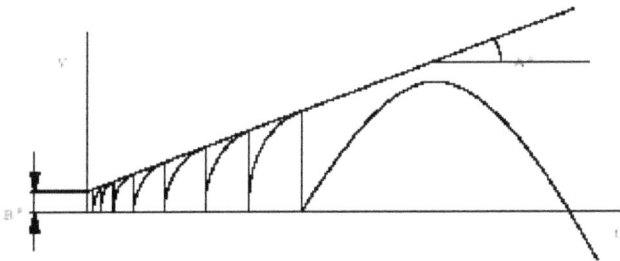

Straight line model of U_b (t) for a VCB.

Figure-76

Winding Clearances

In the initial stages of the design it is important to estimate the clearances required to allow the unit to pass all the prescribed tests without failure. This requires that the voltages between all the windings and to earth are pre-determined from the required test specification. By applying the allowable stress values to these voltages, minimum clearances may be determined for each space. The tables of voltages as previously shown may be used to determine these minimum clearances.

As a guide the impulse type voltages are less onerous than the power frequency voltages by an approximate factor of three. The clearances are not only a function of the type or duration of the voltage but also the shape of the electrodes or the shape of the space between the windings. The winding spaces have three basic major insulation structures;

1 – Radial insulation – which is fitted between concentric windings.

2 – Axial insulation – which is fitted at the ends of the windings to the top and bottom yokes of the core.

3 – Inter-phase insulation – which is essentially between parallel cylinders.

The minor insulation structure is usually restricted to the internal winding insulation; which consists of sticks and stampings, and caps and angles.

Processing

In order to get the best safety factors from oil impregnated insulation, it is necessary to ensure that the moisture level in the insulation is very low at the start of its useful life.

Cellulose insulation as supplied dry from the factory will absorb moisture from the air and as it does so it will grow in size. As the winding and assembly process takes some time this moisture absorption can increase the insulation thickness by approximately six percent. The windings may therefore be bigger than the final design size. This moisture must be removed and the windings shrunk to their final design dimensions before assembling the windings and the core.

The moisture may be removed by a combination of heat and pressure reduction by using vacuum pumps. The windings are simultaneously subjected to a vertical load which consolidates the insulation and removes any slackness and voids.

There are various method used to input heat into the insulation in order to remove the moisture. The most recent method adopted by many manufacturers is heating by using a vapour phase process. This process involves introducing hot kerosene vapour into an autoclave where the core and winding insulation has been isolated. The autoclave is held at a low pressure and the hot kerosene vapour contacts the winding insulation. During the changing phase from vapour to liquid, it gives up heat to the insulation and under the combination of heat and low pressure, the water drains off. The mixture of water and kerosene is then collected in the base of the auto clave and separated. The kerosene is then extracted from the water and recycled through the plant. When the water extraction rate falls to the end point, the autoclave is then isolated and the pressure in the autoclave is lowered to vacuum level and all the remaining surface moisture and kerosene is drawn off. The vacuum in the autoclave is then released with dry air and the core and winding is then removed and a final check is made on the winding clamping pressure and any slack taken up, before fitting the unit into the tank. Each manufacturer will have their own criteria for processing end points but typically values for temperature, vacuum and load pressure are 120 °C, 0.1 mBar, and 5 Mpa, are used until the water extraction rate is reduced to an acceptable level.

The fitting of components such as bushings and tapchangers is then completed quickly under dry air conditions to reduce the amount of moisture absorption during this period. The tank is then sealed and a full vacuum applied for twenty four hours to remove any surface moisture, the unit is then filled with hot degassed oil and left to stand for approximately twenty four hours before any high voltage tests are applied.

Any external bolt on items such as tapchangers and coolers, may require hot oil circulation before connecting to the main tank oil.

Predicting performance

The design of power transformers for specific purposes is an iterative process which involves three stages, initial assumptions for the specification, preliminary design to determine all the basic parameters and the extended performance estimations from these basic parameters.

These performance calculations will later be verified by tests on a manufactured machine, but until the unit has been manufactured and tested, the performance is only an estimated value based on calculations and historical evidence from previously built similar machines.

Most manufacturers will have a Tender design program which will use input from the customers sparse specification and produce an output which will give basic parameters, such distribution of the active material, losses and impedance, cooling requirements and generally enough information to estimate the material weights and dimensions and a budgetary cost of the unit.

The thermal performance is related to load, tap position, winding gradient, hotspot factor and ambient. If the gradient and mean oil rise are known for a type of cooling and a given load, then for a reasonable range, the gradients and mean oil rises can be computed for other loads, taking the indices from the IEC specification.

If some simplifying assumptions are made such as the no-load loss is constant for constant flux regulation, and the tapping range is linear, then the load losses and winding gradients may be estimated for the most onerous tap position and the corresponding mean oil rise determined. These may be calculated for a range of PU loads and a Thermal performance curve can be generated for all types of cooling. In this case it is assumed that plate type radiators with options of additional cooling fans and oil pumps are available, and some estimation can be made for the cooler control temperatures based on the estimated hotspot temperatures.

Testing

The testing of a large power transformer is the final stage in a contract prior to the unit leaving the manufacturer's works. It is the last opportunity to resolve any manufacturing issues that have an effect on the specified performance.

The order in which the final tests are carried out will be detailed in the specification. There are however other tests that are carried out during the manufacturing stage. If there are any uncertainties relating to the ultimate transformer performance, they should be addressed immediately. This allows any corrections or modifications to be carried out as soon as possible, which in turn keeps the cost to a minimum.

Some components may be purchased and the testing may be outside the control if the factory, such as bushings, tapchangers, CT's, Control cabinets etc. In these cases the components should be tested in advance before fitting to the transformer. There are tests associated with the main active parts such as copper and steel which verify that the material is of the correct type and is sound. These extra tests may apply to multiple strip conductors to ensure that there are no shorts between parallel conductors prior to winding.

It is not usually possible to test large transformers at their full rating, and the tests are carried out in two modes. The first mode consists of tests carried out with the unit on no-load, IE the tests are open circuit tests, during which the full voltage is applied and only magnetizing current is required. The second mode is carried out with the transformer short-circuited and full rated current is applied. In this case the supply voltage is only required to overcome the internal impedance of the transformer.

During the initial contract stage it is desirable to agree a test procedure, whereby all the tests required by the contract are listed and a detailed procedure is drawn up as to how these tests will be performed. This is a good discipline as it encourages the manufacturer to ensure in advance that the required test plant is available and suitable to meet the various voltages and currents. It also helps the planning to allow sufficient time to complete the tests whilst determining a time to completion for the contract.

A typical test report is very useful in order to review the test program and may include, but not be limited to;

- Ratio and phase relation check
- Resistance measurements
- No-load losses and excitation current measured at 90%, 100%, and 110 % of the rated voltage at rated frequency. The Harmonics on the magnetizing current are measured at the 100% value.
- Noise level measurements at 100% voltage with and without cooling fans operating.
- Zero-sequence impedance measured on normal and extreme taps.
- Load loss and Impedance measurements.
- Oil samples for DGA analysis.
- Temperature rise tests for all ratings.
- Oil samples for DGA analysis.

- Lightning Impulse tests.
- Applied voltage test
- Induced voltage test
- Oil samples for DGA analysis
- Insulation power factor tests
- Megger tests for core
- FRA measurements
- Control equipment tests including LTC operation, Current transformers, Bushings, Pump and Fan losses.

Test equipment

In order to perform the tests listed some consideration should be given to the test equipment required. It is important to manufacturers that they are able, not only to build a transformer to the customers' requirements but that they also have the necessary test plant to perform all the above tests in accordance with the specifications. Some comments relating to each test or measurement are given.

Ratio test – The specification will call for certain voltages under no-load conditions, the design engineer will convert these voltages to turns, and will fix a value of volts-per-turn, which will result in a design turns-ratio that will be manufactured and measured. The voltage ratio and the turn's ratio may differ in as much as there is a restriction that an integer number of turns are used for each winding. There is a requirement that the actual voltage ratio controlled by the number of turns, should be within 0.5% of the required voltage ratio as specified.

In a large transformer with a large number of small steps in the tapping winding, often requiring tap steps of 0.5%, the design engineer will quite often be limited to tapping steps of a few turns. This in turn will demand a limited range for the volts-per-turn, and discrete values with a 20% difference will be quite common. Measuring the ratio therefore requires accurate instruments that have a tolerance of 0.01% and special ratio-meters are preferred. These will use a bridge system with a null detector to compare actual voltage ratios with known calibrated ratio.

Resistance measurement – The DC resistance of a transformer winding can vary from a few ohms for the HV to a few milliohms for the LV. These windings will have a high inductance and there will be a corresponding time constant whilst the field builds up and the current stabilizes. The equation for the voltage is shown as;

$$V = ir + L\frac{di}{dt}$$

When there is no change in current, the resistance can be taken as V/i.

It is important to measure the temperature of the winding at this stage, as it will be necessary to convert to the operating temperature, most specification require the resistance and corresponding load losses at 75°C.

Care should be taken when the resistance has been measured and the DC source is then disconnected, as a high voltage can be generated when the field collapses.

Loss measurement – The No-load loss often referred to as the Iron loss, will require a supply at rated voltage and frequency. As the unit will be under open circuit conditions, the supply current will be very small and only consist of the magnetizing current. The equipment required to measure this loss, will have to be suitable for all the range of transformers built. This may include units at 50Hz and 60 Hz. The supply voltage will also require readings at plus and minus 10% of the rated value.

The load loss or copper loss will require a supply at rated current, as the unit will be operating under short circuit conditions, the voltage required to supply rated current will only be the impedance voltage. The power factor will be very low and the instruments required to measure the losses must be suitable to measure volts, amps and watts at these low power factors. The ratings of the equipment will be controlled by the most onerous transformers under manufacture, including the highest LV circuit voltage, the highest rated HV current, the range of frequencies and the highest impedance voltage.

The test set up will consist of a motor-generator set, a matching transformer and a bank of capacitors to improve the power factor. This set up would also be used to perform temperature rise tests, where the full input losses will be supplied.

Noise level measurements – The equipment required to measure the noise level is the same as the set up to measure the no-load loss, and the test is often conducted at the same time with the same set up. The actual noise meters will also be required, and the test conducted as specified.

Dielectric tests

Impulse testing

Impulse test are required to verify that the transformer will withstand the effect of lightning hitting the transmission line and travelling towards the transformer. Some form of lightning arresters will restrict the voltage level of the line. These will limit the level of lightning voltage reaching the transformer, and the insulation coordination of the system protection will dictate Basic Insulation Level (BIL) for which the transformer is designed to withstand.

The types of impulse that occur in the system are represented by three basic waves.

- Full wave, which represents a voltage level that has not been interrupted by the protection system.
- Chopped wave, which represents an impulse that has been controlled by the protection system, but after the initial peak.
- Front of wave, where the protection system has interrupted the impulse before it has reached its full potential.
- Switching Impulse, which is of a lesser value than the protection system but is due to switching in the system and has a longer duration than the lightning impulse.

These impulse voltages are artificially produced by a high voltage impulse generator. The operation of this generator is based on the principle that a number of capacitors are charged up in parallel and discharged in series through the transformer. See Figure -77,

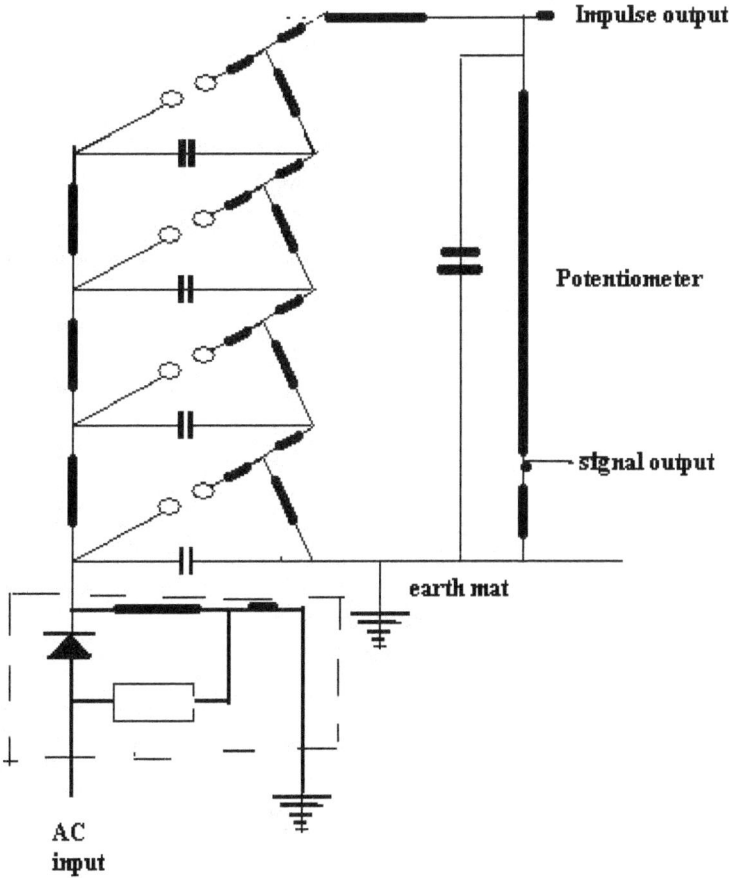

Figure-77

The stage capacitors are normally capable of 200 kVp, and these are charged from a charging unit, which is basically a rectifier circuit. The capacitors are charged up in parallel to the required stage voltage. The sphere gaps are usually set with a gap that will withstand the stage voltage, and by using a fine controller, the gaps are closed until flashover occurs. An alternative method may be used to trigger the gaps by introducing a spark plug in the first gap; this then triggers the other gaps. The flashover of the gaps essentially operates as a switch and reconnects all the capacitors in series. For N stages charged to V volts the output voltage then raises to NV, The effective output voltage is applied to the Transformer winding, the final wave-shape will depend on the load that the transformer presents to the generator and there will be some internal loss due to the load current. The wave shape can be adjusted by altering the resistors within the generator.

International agreement has set target wave shapes that represent these impulses, and these target wave shapes are shown in Figure-78. In reality it is not possible to reproduce these exactly and some tolerance is required, especially when the transformer load is a low voltage winding.

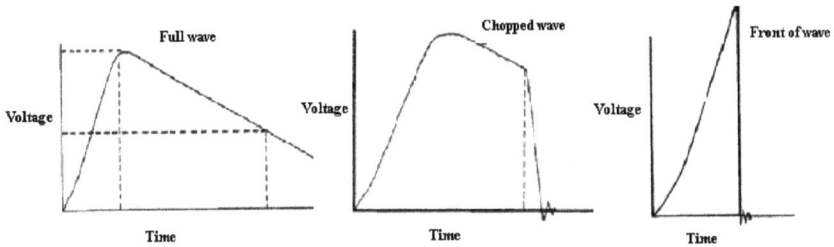

Figure-78

The Impulse test essentially consists of a series of impulses applied to the winding under test. Oscillograms of the voltage and current in the winding are compared and differences between two impulse Oscillograms in the series can identify different types of failure within the test circuit. Further analysis of the Oscillograms can be found in IEC 60076-4 'Guide to Impulse Testing'

Induced Over-potential Test

The induced over-potential tests are carried out by using a sinusoidal voltage which induces a high voltage between the turns, and therefore tests all the winding insulation. The level of this voltage will be at least twice the normal volts per turn and the measurements will be made at the line end bushings, via a potential divider. In order to achieve the high voltages without over-fluxing the core, a high frequency is required, often three times the rated frequency. The test is induced for a number of cycles corresponding to the power frequency applied for one minute. That is for a unit rated at 50 Hz, which has a power frequency of 100 Hz, the number of cycles required would be 6000, or 150 Hz for 40 seconds. The level of voltage at the line terminal will be specified.

Figure-79 shows examples of voltage levels for the 1 minute test

Norm Sys kV RMS	Highest sys kV RMS	BIL kVp 1/50 microsecond	Over Potential kV RMS
66	72.5	350	140
132	145	550	230
275	300	1050	460
400	420	1425	630

Figure-79

Note that the volts per turn for the Over Potential test could be as high as 630 * 400/√3 = 2.73 times the normal volts per turn. Therefore in order to prevent over fluxing, the frequency of the supply would have to be in the order of 2.73 times the rated frequency.

If the over-potential test is carried out by using a three-phase supply, then the voltage between line terminals would be 630*√3 = 1091 kV, If the test is carried out as three single phase tests, then the possible voltage between the line terminals is 630*1.5 = 945 kV. This assumes a three-limb core where the return flux is 50% in each of the two untested limbs.

At these high voltages between terminals for 6000 cycles, it was taken that this test would be more onerous than a Switching Surge test level which would only be applied a few times. The Switching Surge test level for a 1425 BIL would be 1050 kVp to neutral or 1575 kVp between line terminals. The strength of insulation under impulse conditions may be taken as at least twice that of the power frequency tests.

Partial Discharge Test

In a high voltage transformer which has a complex insulation structure containing a mixture of cellulose, moisture, particles and oil; the different dielectric properties, which are also a function of temperature; create local voltage stresses in each property which may exceed the electric strength. In this case there will be a partial breakdown which will take the form of an electric discharge. This discharge, however small may cause permanent damage to the insulation and if subsequent breakdowns occur, the effect may be cumulative and eventually create a total breakdown in the form of a complete arc.

During the high voltage power frequency dielectric test, using special detection equipment, a measure of these partial discharges may be assessed. The IEC 60076-3 has described a method to measure these partial discharges, and has defined an acceptable limit in association with test voltage duration. Measurements are taken before and after the highest test level and compared to ensure that there is no permanent damage due to the test level. That is there is no hysteresis effect.

PART 2

REACTORS

General

There are two basic forms of reactor construction. One is the gapped-core construction and the other is air cored and they are usually of the magnetically shielded type.

The choice between the two types in any given instant depends on the operational requirements. In most cases a slightly 'drooping' voltage/current characteristic (i.e. a reduction of inductance with increasing supply voltage) is an advantage, these characteristiscs are a function of the gapped core and iron flux paths. In other cases a linear characteristic is required up to a voltage as high as 150% of normal.

Provided that some degree of non-linearity is acceptable above the normal rated system voltage, the gapped-core reactor will be preferred; it makes better use of the allowable winding height. In practice the majority of shunt reactors are nowadays of gapped-core construction whilst the series reactor, which rerquire a more linear voltage/current characteristic are normally air cored and magnetically shielded; the voltage across the winding is small and therfore allows for a shorter winding height.

With Air cored reactors, in most cases, the core is actually oil.

Characteristics for reactors

Y axis – Magnetic flux linkage peak value (p.u.)

X axis – Current value (p.u.)

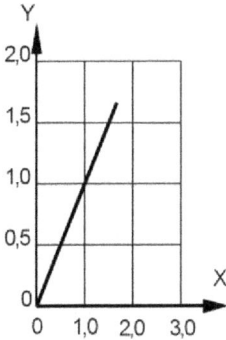 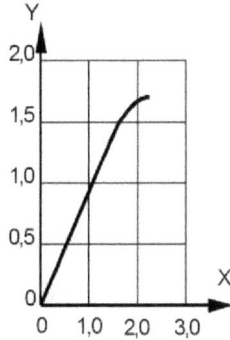 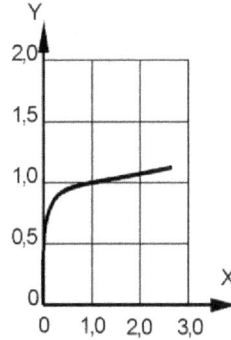

Linear   Non-Linear   Saturable

- Linear – air path

- Non-Linear – air + iron path

- Saturable – Iron path

Self-Inductance of a coil

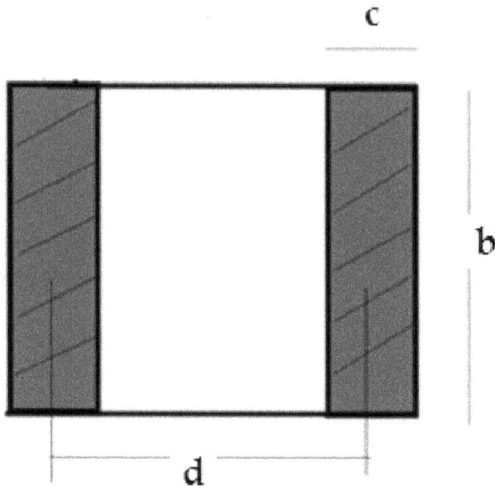

$$L = K\ (d/25.4)\ N^2\ 10^{-6}\ \text{milli-henries}$$

Where;

K is a function of the coil shape and is related to the ratio b/d for different values of c/d.

d is the mean diameter in mm.

N is the number of turns in the coil.

shape factor K

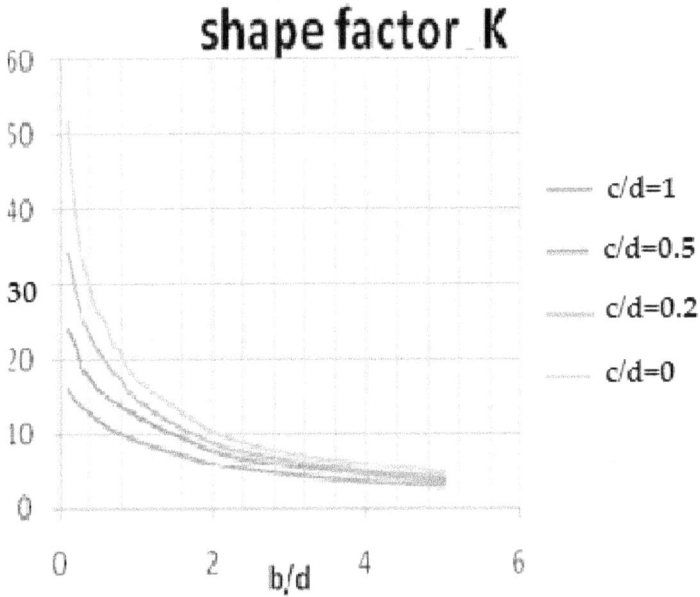

If the coil is placed inside a magnetic shield, the shape factor is modified depending on the clearance to the shield.

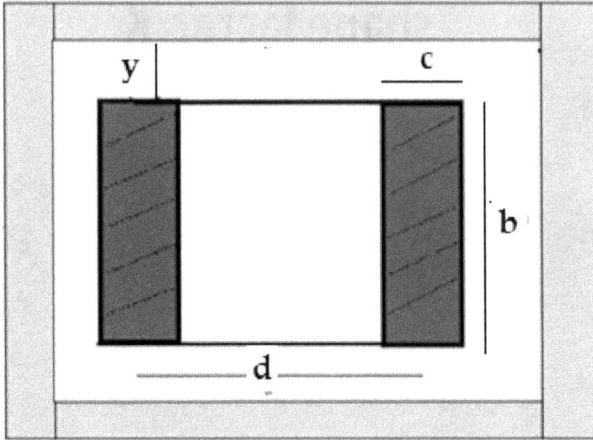

Yoke Clearance at c/d = 0.125

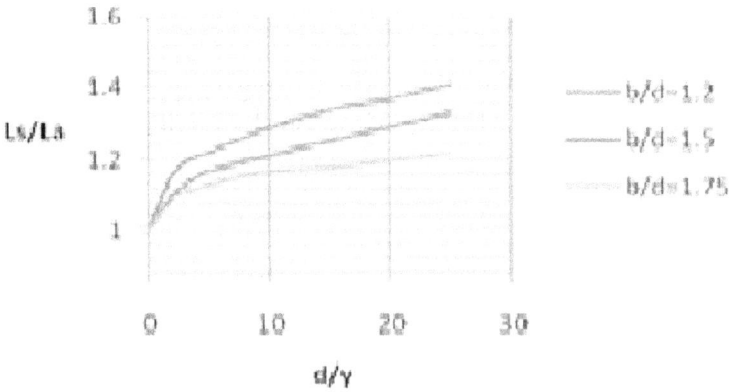

Ls = Inductance in shield
La = Inductance in air

Series Reactors

Example -1

b = 480

c =86.5

d =460

y =50

b/d = 1.075

c/d = 0.193

K= 13.8

N = 77 turns

L =1.482 mH

For coil in shield,

d/y =9.2

La/Ls = 1.29

K=17.8

L 1.91 mH

Example-1 shows the output from an EXCEL spread sheet, that has been set up to produce a preliminary design for a single-phase shielded reactor which will present a reactance of 0.542 ohms, when inserted in series with the output of a 13.8 kV transformer winding, normally a tertiary to restrict fault currents. The preliminary input data for the spread sheet are shown in grey, calculations from this data are shown in black.

1-ph Series 13.8 kV Tert Reactor	Estimared dimensions						
Required ohms	0.542						
KVAr	366.9	Phases	1	Frequency	60	Cooling	ONAN
voltage drop	392.6	Coil length (b)	480	Numb legs	2		
line Amp	724.6	Coil mean dia (d)	446	total leg lth	1160		
Tert BIL	150	Coil width (c)	86.1	total yoke lth	1548.2		
		Turns	77	Shield width	553		
Wdg Rise	65	c/d	0.193				
Shied depth mm	100	b/d	1.076				
Shield area sqcm	536.41	Ka	13.8	inductance H	0.001437		
flux density	0.23	d/y	8.92				
Flux Wb	0.0247	Yoke clearanceT/B	50	ohms	0.542		
Volts per turn	6.5734	Ks	17.8	amps	724.6	In Shield	561.8
		kg		w/kg	kW	va/kg	kva

yoke kg/cc	0.00765	635.31	0.09	0.0572			
leg kg/cc	0.00765	476.01	0.09	0.0428			
Total kg		1111.32		0.1000			
		Henries = K*d*N^2*10^-6 mH			0.001		
Winding		Reactance = 44/7*f*L			0.54		
Type	Disc						
SF	0.5						
Tert KV	13.8					b/d	1.076
Turns Tot	77					c/d	0.193
Turns Axial	28					d/y	8.92
Turns Radial	3						
Amps	724.6					Ka	13.8
Cond bare	13 x 3.5 -7//	7	13	3.5		Ks	17.8
Cond Ins	0.5 pap +0.1 en	0.5		0.1		Ks/Ka	1.290
Cond Covd	AXIAL	13.5		RADIA	4.1		
Cond area	318.5			volts in air at	724.6	392.614	
A/mm2	2.275			volts in shield at	724.6	506.415	
Stamp thk	4			Volts due to shield at	724.6	113.801	
STMP/CIRCLE	16	STMP WIDTH	30				
Wdg height	480	nominal	859.6				
Ins T/B	50			at 2 tesla amps =	6300.87		
Wdg Rad	86.1			Max s/c amps =	8127.21		
Wdg ID	359.9	Wdg OD	532.1	s/c votls =	5393.19		
MT	1.40	Cooling area sqcm	40504				
Length/leg	107.93			X sc =	0.664		
Tot wt/Trans	307.10			X fl =	0.699		
Res @75	0.007354			X fl/Xsc =	1.053		
I2R/leg	3860.98						
STRAY LOSS %	64.0						
Tot loss/leg	6330.3						
w/cm2	0.156						
Oil Passes	1.4						
Gradient	20.2						
Major Ins	100						
Test kV	250						
vertical Ducts	8						
Clearance to shield	21						
Losses @ 75	Calc	Guar	Test	Imp %	Calc	Guar	Test
Wdg I2R							
Conns							
Wdg Stray							
Tank & Clamps							
Winding kW	6.3303						
Shield kW	0.1000						
Total kW	6.4303						

13.8 kV 1-phase Tertiary Reactor

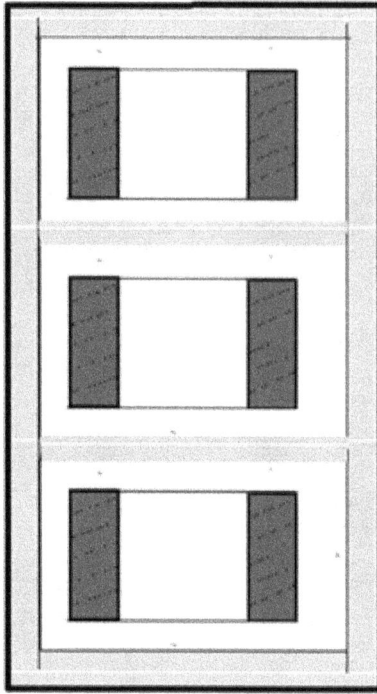

Three phase unit aligned vertically

Example -2 presents the initial design calculations for a three phase unit mounted vertically which may be fitted inside the main tank of a three phase auto transformer. This will also reduce the short circuit current in the tertiary; which is 1.46 ohms per phase in the 22 kV tertiary winding.

Series tertiary Reactor 3-ph 22 kV	Estimared dimensions						
ohms/ph	1.46						
KVAr	867.6	Phases	3	Frequency	50	Cooling	ONAN
voltage drop	523.4	Coil length (b)	650	Numb legs	6		
line Amp	455	Coil mean dia (d)	651	total leg lth	5244		
Tert BIL	150	Coil width (c)	103	total yoke lth	4320		
		Turns	101	Shield width	585		
Wdg Rise	65	c/d	0.158				
Shied depth mm	100	b/d	0.998				
Shield area sqcm	567.45	Ka	14	Inductance H	0.003660		
flux density	0.266	d/y	5.8125				
Flux Wb	0.0302	Yoke clearanceT/B	112	ohms	1.150		
Volts per turn	6.7018	Ks	17	amps	455	In Shield	374.7
		kg		w/kg	kW	va/kg	kva
yoke kg/cc	0.00765	1819.05	0.1	0.1819			
leg kg/cc	0.00765	2208.12	0.1	0.2208			
Total kg		4027.17		0.4027			
		Henries = K*d*N^2*10^-6 mH		Target	0.00366		
Winding		Reactance = 44/7*f*L		Target	1.15		
Type	Disc						
SF	0.5						
Tert KV	22					b/d	0.998
Turns Tot	101					c/d	0.158
Turns Axial	54					d/y	5.8125
Turns Radial	2						
Amps	455					Ka	14
Cond bare	3.6 x 1.8 -23//	46	3.6	1.8	Ks	17	
Cond Ins	0.85	0.75	0.1		Ks/Ka	1.214	
Cond Covd	AXIAL	7.95	RADIAL	2.65			
Cond area	283.176		volts in air at	455	523.425		
A/mm2	1.607		volts in shield at	455	635.587		
Stamp thk	4		Volts due to shield at	455	112.162		
STMP/CIRCLE	25	STMP WIDTH	30				
Wdg height	650	nominal	656.1				
Ins T/B	112		at 2 tesla amps =	3421.05			
Wdg Rad	103		Max s/c amps =	8723.68			
Wdg ID	548	Wdg OD	754	s/c votls =	10878.91		
MT	2.05	Cooling area sqcm	226873				
Length/leg	206.65		X sc =	1.247			
Tot wt/Trans	1568.26		X fl =	1.397			
Res @75	0.015835		X fl/Xsc =	1.120			
I2R/leg	3278.33						
STRAY LOSS %	12.5						
Tot loss/leg	3689.8						
w/cm2	0.016						
Oil Passes	2.7						
Gradient	2.1						
Major Ins	100						
Test kV	250						
vertical Ducts	8						
Clearance to shield	63						
Losses @ 75	Calc	Guar	Test	Imp %	Calc	Guar	Test
Wdg I2R	9.83						
Conns	1.50						
Wdg Stray	1.23						
Tank & Clamps	3.00						
Load loss KW	15.57						
Shield kW	0.40						
Total kW	15.97						

22kV 3-phase Tertiary Reactor

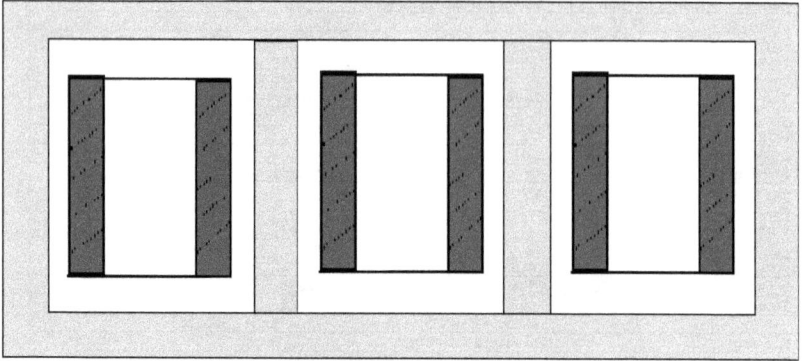

Example – 3 Three Phase reactor in series with a 400 kV transmission line.

Series Reactor 3-ph 750 MVA		Estimared dimensions					
ohms/ph	16.5						
KVAr	86866.3	Phases	3	Frequency	50	Cooling	ONAN
voltage drop	18381.2	Coil length (b)	1625	Numb legs	4		
line Amp	1083	Coil mean dia (d)	1098	total leg lth	11284		
Winding BIL	1425	Coil width (c)	390	total yoke lth	13388		
		Turns	337	Shield width	1060		
Wdg Rise	55	c/d	0.355				
Shied depth mm	160	b/d	1.480				
Shield area sqcm	1628.16	Ka	11	inductance H	0.054003		
flux density	0.731	d/y	2.506849				
Flux Wb	0.2380	Yoke clearanceT/B	438	ohms	16.972		
Volts per turn	52.8442	Ks	16	amps	1083	In Shield	744.6
		kg		w/kg	kW	va/kg	kva

yoke kg/cc	0.00765	16175.06	0.22	3.5585			
leg kg/cc	0.00765	1363.31	0.22	0.2999			
Total kg		17538.37		3.8584			
		Henries = K*d*N^2*10^-6 mH				Ta	0.05400
Winding		Reactance = 44/7*f*L				Ta	16.97
Type	Disc						
SF	0.5						
System KV	400					b/d	1.480
Turns Tot	337					c/d	0.355
Turns Axial	90					d/y	2.50685
Turns Radial	4					Ka	11
Amps	1083					Ks	16
Cond bare	5.25 x 1.3 -27//	54	5.25	1.3		Ks/Ka	1.455
Cond Ins	2.76	2.5	0.26				
Cond Covd	AXIAL	13.78	RADIAL	23.56			
Cond area	700.245	13.52	volts in air at		1083	18381.2	
A/mm2	1.547		volts in shield at		1083	26736.31	
Stamp thk	4		Volts due to shield at		1083	8355.098	
STMP/CIRCLE	28	STMP WIDTH	40				
Wdg height	1625	nominal	1576.8				
Ins T/B	438		at 2 tesla amps =		2963.06		
Wdg Rad	390		Max s/c amps =		7555.81		
Wdg ID	708	Wdg OD	1488	s/c votls =	151100.4		
MT	3.45	Cooling area sqcm	2487937				
Length/leg	1162.94		X sc =		19.998		
Tot wt/Trans	21824.37		X fl =		24.687		
Res @75	0.036038		X fl/Xsc =		1.234		
I2R/leg	42269.15						
STRAY LOSS %	71.3						
Tot loss/leg	72398.7						
w/cm2	0.029						
Oil Passes	4.5						
Gradient	3.8						
Major Ins	100						
Test kV	250						
vertical Ducts	8						
WDG to shield mm	265						
Losses @ 75	Calc	Guar	Test	Imp %	Calc	Guar	Test
Wdg I2R	126.81						
Conns	1.50						
Wdg Stray	90.39						
Tank & Clamps	3.00						
Load loss KW	221.70						
Shield kW	3.86						
Total kW	225.55						
Max kW							
Cooler kW							

750 MVA Throughput Series Reactor

High voltage Series Reactors

Purpose
They are introduced into a high voltage line to limit the fault current during short circuits. They may also be used to provide a connection between two large high voltage systems, in order to reduce the fault currents which potentially can be transferred between the two systems.

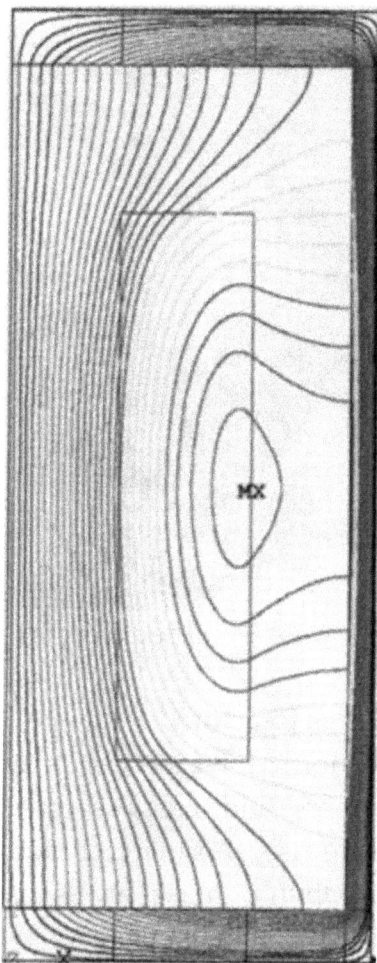

Flux in shielded reactor

4-limb Shielded Reactor

High Voltage Series Reactor

Apart from the normal requirements of all large high voltage transformers and reactors (insulation strength, losses, temperature rise ,short circuit mechanical strength, over loading capability, noise levels etc.), a particular featue in the design of a series reactor is the need to ensure that it willstand the enormous forces that are imposed in the event of a close up fault on the outlet side when the short circuit MVA on the inlert side is at or near to the maximum system MVA. Virtually the whole of the system fault MVA energy is stored in the magnetic field of the reactor.

The design of the unit shown is that of a three phase magnetically shielded air cored reactor, there are no core laminations within the coils, the core is of a four limbed magnetic shielded type within which the coils have to be securely located.

Forces in an ideal winding

The forces generated in a single symetrical winding by the current passing through it consists of an axial crushing force and a radial bursting force tending to flaten the coil and increase the diameter, so as to increase the self inductance. The axial forces are relatively easy to contain provided that the insulation is strong enough, securely located and dried out sufficiently. Epoxy bonding the CTC prevents crushing and tilting of the individual conductor strands. The hoop stress in the winding conductors caused by the radial bursting force, must be well within the permissible limit of the critical limit for the CTC used in the winding.

Details of Application

The China Light and Power Company's 400kV system in the New Territories of Hong Kong connects the Castle Peak Power Station complex with the main load center of Kowloon, some 28 km distance, with interconnection tap-off points for bulk infeed to various major load centers.

There are two double circuit lines from the power station to the outskirts of Kowloon – one route being more or less direct and the other a circuitous and longer route through the heart of the new territories. The two double circuits complete a ring to the outskirts of Kowloon and power is delivered to Tai Wan substation, situated towards the tip of the Kowloon peninsula, from two substations by 700 MVA 400 kV cables. Following the installation of two reactors there will be a double circuit from one substation and a single circuit from the other with provisions for a second circuit later.

The difference in overall route length and the complex characteristics of the system inductance, cable capacitance, shunt reactance and capacitance have necessitated the provision of series reactors to control load sharing between the parallel circuits into Kowloon, and also to provide some degree of short circuit fault limitation.

Studies were made to determine the dynamic stability of the system and the characteristics of the reactors were thereby determined. In summary these are;

Voltage   400kV nominal   420 kV rated
Throughput 750 MVA nominal   787.5 MVA rated
Current   1083 Amps   1083 Amps
Impedance at nominal throughput 7.5% of 400kV on 750 MVA
Impedance at rated throughput 7.143% of 400 kV on 787.5 MVA
Impedance at short circuit current of 14kA 6.43% of 420 kV referred to 787.5 MVA
Impedance tolerance ± 7%
Cooling ONAN with directed oil flow
Overvoltage capability 440kV for 0.5 hours
Overload capability 2.05 p.u. for 0.5 hours
Lightning impulse test 1425 kVp
Switching Impulse test 1050 kVp
Inducted overvoltage test 630 kV RMS
Noise level 82 dBA

The above and other characteristics were all tested with the relevant IEC specifications. For the record the very comprehensive test program proved that the reactors met all the requirements of the purchaser's specification.

Shunt Reactors

Shunt Reactors

Shunt reactors for a given nominal system voltage are designed for continuous operation at the recognised associated highest system voltage, whilst remaining within temperature rise. For example a 100MVAr, 275 kV unit can be operated continuously at 302.5 kV at which voltage its rating is 121 MVAr.

Purpose

To prevent over-voltages on the lines when they are lightly loaded

- Lightly loaded lines present a capacitive load and high voltages can result, if the lines are loaded with inductances, this compensates the capacitive current and the voltages are stabilized.

- This is very desirable when the lines join two different systems across international boundaries.

A simple shunt inductance is required

- Normally a single winding for each phase

- 3-ph or 3x1-ph units are normal.

- Normally only required for HV lines above 132 kV.

- If the inductive current required is I and the phase voltage is V, the reactance per phase is given by;

- $X = V/I$

- If the frequency is f then the inductance required is given by;

- $L = X/2\pi f$

The general arrangement for one leg is shown. The core packets are separated with ceramic spacers, and for high voltages a copper screen may be placed to reduce the electric field at the sharp corners of the packets.

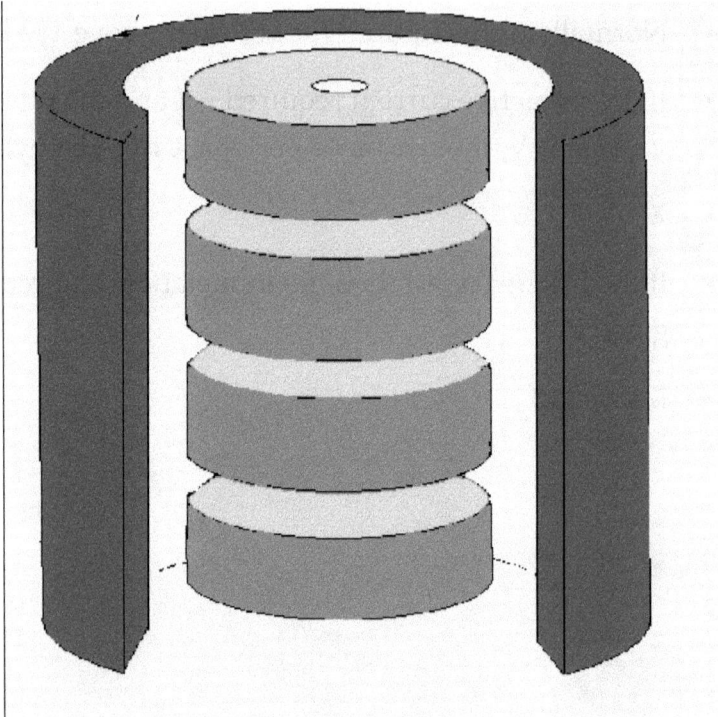

General arrangement of core and coil.

Typical starting sequence for the design follows;

1 Choose a core size, based on previous experience.
2 Choose an initial flux density and a typical value for the ratio of stray flux to core flux (α).
3 Determine the turns per phase based on the required reactance.
4 Determine the effective area of the gap.
5 Determine the flux density in the gap.
6 Determine the required total gap length.
7 Detail the gaps.

Typical Diameters for gapped core reactors.

Core diameter mm	Location washer mm	Arrangement of holes	Number of spacers	Core area Sqcm
144/504	144/500	8 on 240 pcd 16 on 410 pcd	24	1640
160/580	160/556	9 on 270 pcd 18 on 450 pcd	26	2185
192/672	192/668	10 on 310 pcd 20 on 550 pcd	30	2915
216/756	216/752	12 0n 350 pcd 24 on 620 pcd	36	3690
216/854	216/880	10 on 330 pcd 20 on 540 pcd 20 on 750 pcd	50	4799

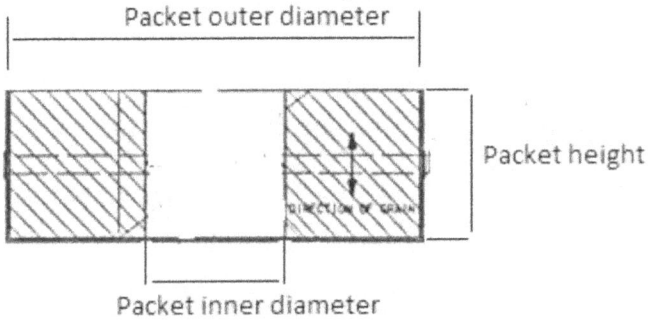

Packet outer diameter

Packet height

DIRECTION OF GRAIN

Packet inner diameter

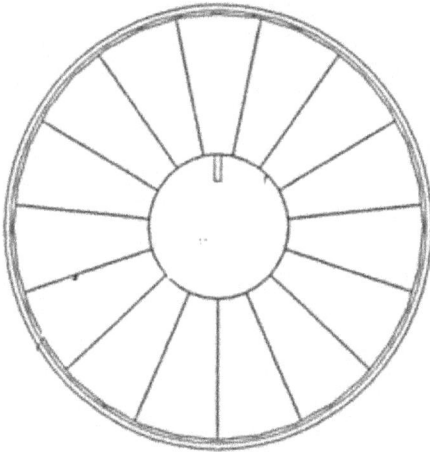

Arrangement of segments

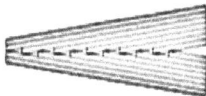

Typical arrangement of laminations

Physical Design
The dimensions and type of the winding will mainly be
controlled by the BIL (Basic Impulse Level) and the
maximum axial and radial stresses that can be tolerated.

The higher the BIL, the longer the winding has to be.

The magnetic circuit is complex and the reluctance for
each part must be assessed and then added to estimate a
composite reluctance for normal and maximum voltages.

General arrangement of leg

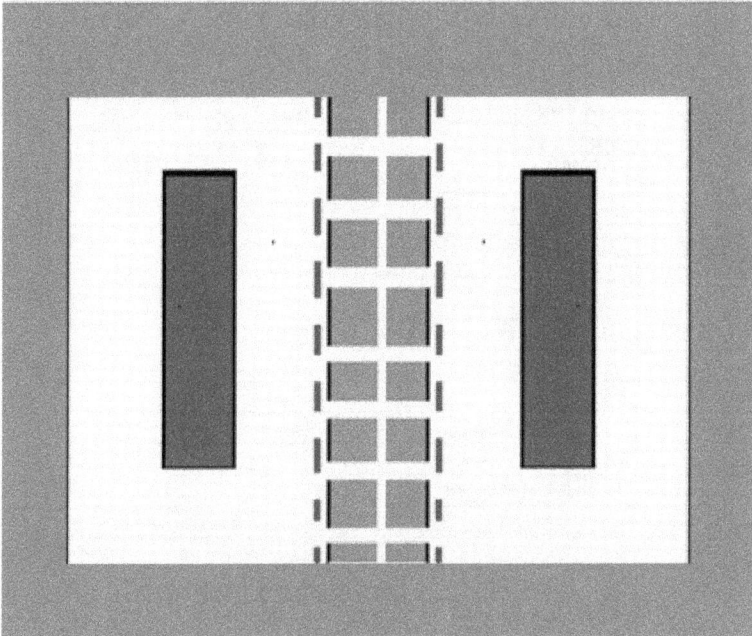

Electrostatic screen for high voltages

0.128 MVAr 16 kV 1-ph Neutral reactor

Input data		Calc data		Calc data	
MVAr	0.13	Amps	8.00	Wdg area cm²	388.53
KV line	16	Volts/ph	16000.0	discs	101
Freq	50	Gap area cm²	2030.09	Turns / disc	15.92
Core id mm	144	Gap B l/cm²	2035.77	Cond bare T mm	1.40
Core od mm	504	Turns	1615	Cond bare W mm	5.00
Core area cm²	1640	Gap length cm	11.29	Strips/ turn	1.00
Biron l/cm²	2520	Number Gaps	3	Cond axial mm	6.00
alpha-1	0.08	Area oil cm²	1803.35	Cond rad mm	2.40
Spacer thk mm	36	Area w cm²	481.52	Turn csa mm²	6.79
Winding Ht mm	974	Area stray cm²	2284.87	MT m	2.21
Core Wdg mm	79	Stray B l/cm²	236.07	Length m	3561.99
Wdg shield mm	89	Wdg id mm	662	Cond Kg	216.06
Wdg T Yoke m	100	Wdg od mm	742	Resiatance/ph	11.3837
Wdg B Yoke m	153	Wdg rad mm	40	I²R kw	0.73
A/mm²	1.18	Win Ht mm	974.00	Stray %	0.50
FE W/kg	2.4			Cu Loss kw/ph	0.73
Wdg SF	0.28	shield area cm²	869.2	Stamp/circle	20
Stmp Thk mm	3.6	shield width cm	17.25	Stamp width mm	24
		shield length cm	4983.40		
strip + 1p	5x1.4	shield kg	3275	Cooling area cm²	178553
		Number packets	4	w/cm²	0.004
		Packet Ht cm	26.93	Wdg Grad °C	1.62
		leg kg	1381		
		Fe kg	4656		
		Fe kw	11.17	Reactance Ω/ph	2092.70
		load kw	17.4		

16 kV Neutral Reactor

Typical Flux distribution

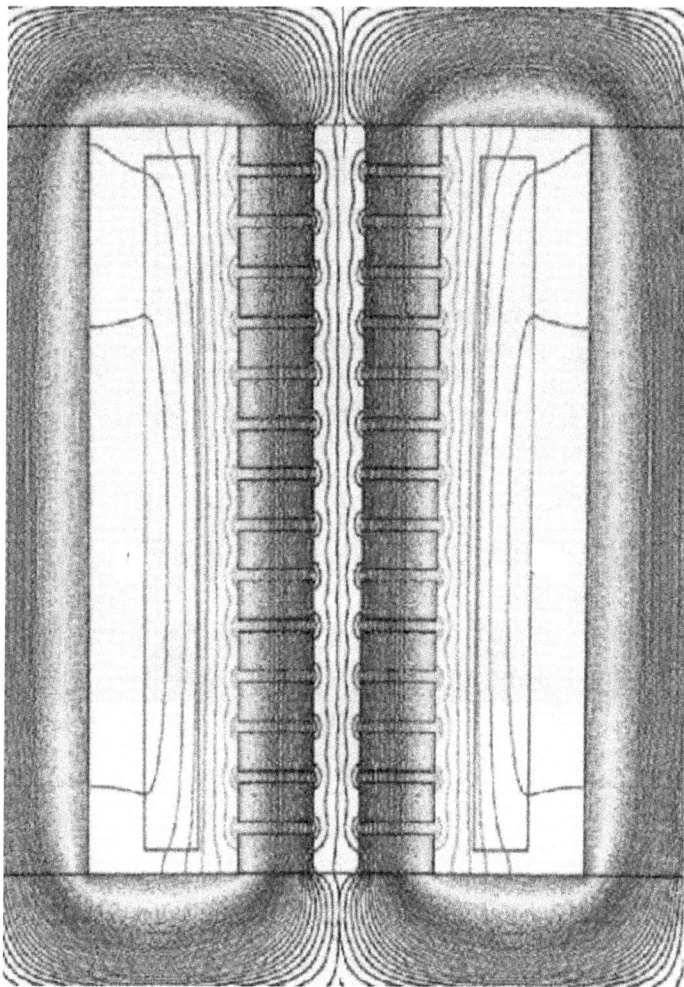

Sample field plot

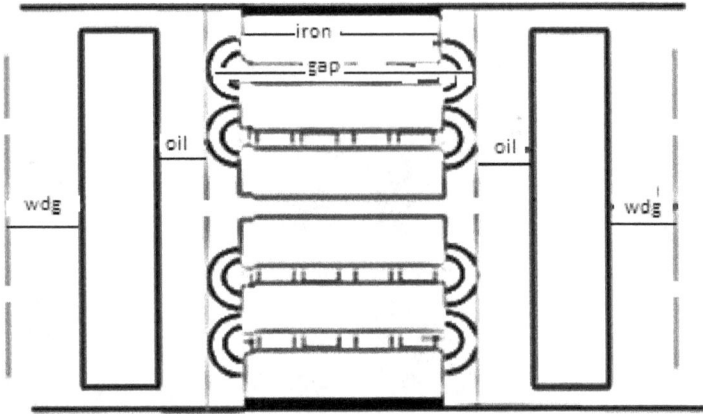

Components of flux distribution

Area iron is the core area containing the flux in the iron.
Area gap is the area of the gap taking account of the
fringing.
Area oil is the area in the oil containing the stray flux
inside the winding
Area wdg is the area containing the stray flux outside the
winding.

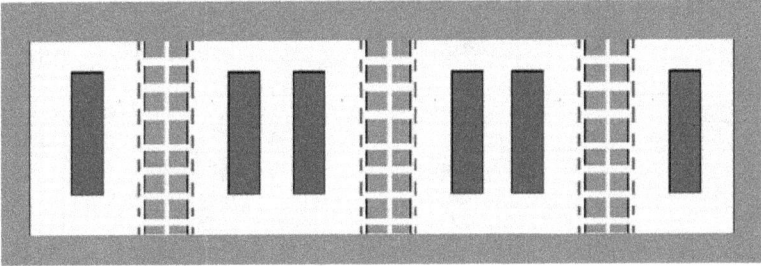

150 MVAr 400 kV unit on test with test oil/air bushings

150 MVAr 400 kv					
Input data		**Calc data**		**Calc data**	
MVAr	150	Amps	216.51	Wdg area cm²	4333.18
KV line	400	Volts/ph	230947	Number discs	132
Freq	50	Gap area cm²	4419.82	Turns/disc	22.54
Core id mm	216	Gap B l/cm²	12255	Cond bare T mm	2.00
Core od mm	756	Turns	1490	Cond bare W mm	11.00
Core area cm²	3696	Gap length cm	46.86	Strips/ turn	4.00
Biron l/cm²	14655	Number Gaps	13	Cond axial mm	13.00
alpha-1	0.289	Area oil cm²	4901	Cond rad mm	4.00
Spacer thk mm	36	Area w cm²	4381	Turn csa mm²	85.36
Winding Ht mm	2380	Area stray cm²	9281	MT m	3.86
Core Wdg mm	145	Stray B l/cm²	2413	Lenght m	5750.89
Wdg Wdg mm	252	Wdg id mm	1046	Cond Kg	13156
Wdg T Yoke mm	77	Wdg od mm	1410	Resiatance/ph	1.4620
		Win Ht mm	2380		
Wdg B Yoke mm	77	Wdg rad mm	182	I²R kw	68.53
Wdg end yoke mm	261	core cent-end yoke cm	97	Tank and clamp kw	30.00
Volts per turn	155.0	Core centres cm	166.21		
A/mm²	2.5	yoke perimeter cm	1696.5	Stray %	23.52
FE W/kg	1.5	Leg length cm	253.4	Cu Loss kw/ph	84.65
Wdg SF	0.2978	Yoke area cm²	2587.2	Stamp/circle	32
Stmp Thk mm	5	Yoke Ht cm	34.22	Stamp width mm	30
strip 11x2-2//-2 ct	22x2+2p	Yoke kg	33578	Cooling area cm²	1858270
		Number packets	14	w/cm²	0.046
		Packet Ht cm	206.54	Wdg Grad °C	17.95
		leg kg	17313	Clamp tons/phase	101.77
		Fe kg	50892		
		Fe kw	76.34	Reactance Ω/ph	1169.16
		load kw	360.3		

Example output for a 150 MVAr 400 kV shunt reactor, supplied to SSEB Torness.

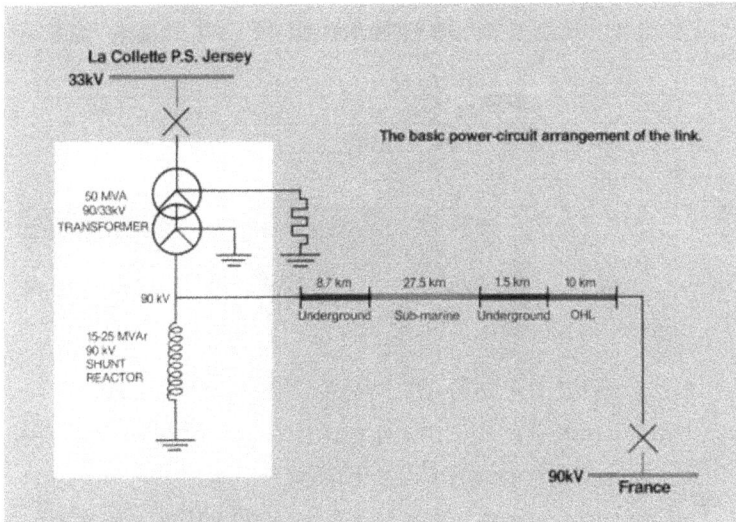

The basic power-circuit arrangement of the link.

The 90 kV cable link between France and Jersey has a Transformer and a Reactor combination as shown.

15-25 MVAr 90 kv min turns						
Input data		Calc data		Calc data		
MVAr	25	Amps	160.38	Wdg area cm²	1071.22
KV line	90	Volts/ph	51963	Number discs	93
Freq	50	Gap area cm²	2476	Turns/disc	15.56
Core id mm	160	Gap B l/cm²	11298	Cond bare T mm	1.80
Core od mm	560	Turns	721	Cond bare W mm	4.25
Core area cm²	2027	Gap length cm	18.23	Strips/ turn	9.00
Biron l/cm²	13800	Number Gaps	5	Cond axial mm	9.80
alpha-1	0.16	Area oil cm²	1714	Cond rad mm	10.75
Spacer thk mm	35	Area w cm²	1319	Turn csa mm²	66.78
Winding Ht mm	1187	Area stray cm²	3033	MT m	2.45
Core Wdg mm	65	Stray B l/cm²	1735	Lenght m	1768.91
Wdg Wdg mm	257	Wdg id mm	690	Cond Kg	3166.05
Wdg T Yoke mm	136	Wdg od mm	870	Resiatance/ph	0.5748
		Win Ht mm	1187		
Wdg B Yoke mm	89	Wdg rad mm	90	I²R kw	14.78
Wdg end yoke mm	261	core cent-end yoke cm	70	Tank and clamp kw	5.00
Volts per turn	72.03504	Core centres cm	112.75		
A/mm²	2.25	yoke perimeter cm	1113	Stray %	12.16
FE W/kg	1.5	Leg length cm	141.2	Cu Loss kw/ph	16.58
Wdg SF	0.48	Yoke area cm²	1418.9	Stamp/circle	20
Stmp Thk mm	3	Yoke Ht cm	25.34	Stamp width mm	30
ctc 4.25x1.8-9//	0.15en +1 p	Yoke kg	12084	Cooling area cm²	410437
		Number packets	6	w/cm²	0.040
		Packet Ht cm	122.97	Wdg Grad °C	15.92
		leg kg	5653	Clamp tons/phase	49.49
		Fe kg	17737		
		Fe kw	26.61	Reactance Ω/ph	331.72
		load kw	81.4		

Example output for a 15-25 MVAr 90 kV reactor; the output is shown for the main winding only. This unit was part of a transformer and reactor combination that was situated in Jersey and connected to a 90 kV submarine cable from France. The 15 MVAr rating is achieved by LTC controlling the number of turns.

Core and windings of 121 MVAr 550 kV 3 phase shunt reactor of the 5 limb core construction. Built to Australian specifications, for a site in Victoria, Australia.

121 MVAr 550 kV				Centre Terminal	
Input data		Calc data		Calc data	
MVAr	121	Amps	127.02	Wdg area cm²	3767.74
KV line	550	Volts/ph	317552	Number discs	147
Freq	50	Gap area cm²	4411.6	Turns/disc	29.38
Core id mm	216	Gap B l/cm²	12198	Cond bare T mm	1.60
Core od mm	756	Turns	2093	Cond bare W mm	8.00
Core area cm²	3696	Gap length cm	38.79	Strips/ turn	4.00
Biron l/cm²	14560	Number Gaps	11	Cond axial mm	10.50
alpha-1	0.27	Area oil cm²	5570.7	Cond rad mm	5.70
Spacer thk mm	35	Area w cm²	4031.7	Turn csa mm²	50.18
Winding Ht mm	2250	Area stray cm²	9602.4	MT m	3.94
Core Wdg mm	165	Stray B l/cm²	2103.2	Lenght m	8245.14
Wdg Wdg mm	324	Wdg id mm	1086	Cond Kg	11087.4
Wdg T Yoke mm	190	Wdg od mm	1421	Resiatance/ph	3.5658
Wdg B Yoke mm	150	Wdg rad mm	167	I²R kw	57.53
Wdg outer leg n	237	Yoke perimeter cm	1610		
A/mm²	2.52	Win Ht mm	2250.0	Stray %	11.26
FE W/kg	1.6	Core centres cm	174.49	Cu Loss kw/ph	64.01
Wdg SF	0.28	shield area cm²	2624.2	Stamp/circle	36
Stmp Thk mm	4.8	shield width cm	36.50	Stamp width mm	30
strip 8x1.6-twin	8X1.6	shield kg	31931	Cooling area cm²	970104
Leg length mm	2590	Number packets	12	w/cm²	0.066
shield length mm	5457.7	Packet Ht cm	220.21	Wdg Grad °C	26.00
volts per turn	151.72	leg kg	18459		
		Fe kg	50390	clamping load Tonnes	100.45
		Fe kw	80.62	Reactance Ω/ph	2706.16
		load kw	276.6	alpha-2	0.37529

Example of a 121 MVAr 550 kV reactor for Australia. The high voltage winding has a centre entry terminal with a BIL of 1550 kVp. The neutral terminals had a BIL of 550 kVp, suitable for single phase auto reclosing with associated neutral reactors.

Example showing the arrangement of packets in a 3-phase
unit with outer limbs.

Design Procedure for Shunt Reactors

There are many things to consider during a design process, and the process is usually iterative. The number of iterations may be reduced by following this sequence of events. The sequence of events is:

Determine the required clearances in oil between;

Winding & Top and bottom yokes.

Winding & Tank.

Winding and Electrostatic Shield.

Clearance between Line & neutral.

(Winding length proportional to BIL).

Select a core diameter (D) based on MVAr (as a rough guide use 500 mm for 50 MVAr and 900 mm for 200 MVAr. See typical values in the table)

Select the reactor voltage (Line kV).

Select the frequency. (50 or 60 hertz).

Select an initial value for alpha. (Use 0.3 to start with).

Select an appropriate flux density in the iron. (Should not be greater than 1.4 Tesla).

Calculate the number of turns.

$(N = V / \sqrt{3}*4.44*f*B*A)$.

Calculate the current in the winding.

$(I = MVAr/\sqrt{3}V)$.

Calculate the effective area for the gap (g).

$Ag=\pi/4*((Do+1.69/\pi*g)^2-(Di-1.69/\pi*g)^2)$

Calculate the total length of air gap

$\Sigma g = 1.78*I*N/Ba$

Select current density (2 A/mm^2 for naturally cooled units).

Select a winding space factor (0.4 for high voltage or layer type windings, 0.6 for low voltage)

Select the clearance between the winding and the core.

(Take care that a stress cylinder may be required)

Select the height of the ceramic spacers.

(Standard = 35 mm, the standard spacer diameter is normally 68 mm).

Calculate the required height of the winding and the radial dimension.

Calculate the winding outside diameter.

Select winding clearances top and bottom to the yokes.

Calculate the core height.

Calculate the number of iron packets.

Calculate the height of each packet.

Calculate the force required for the core clamping.

Force = $B^{2}*A/\mu_0$ - Newton

Winding design

Assume current density 'j' and space factor 'sf', and the product of winding length 'l' and width 'd'.

$l*d=I/j*N*/sf$

Winding Di=core diameter +2*clearance to winding (Check clearance to winding is greater than 2*g)

Select winding length 'l'

Calculate 'd'

Winding Do = winding id+2*d

Leg length = winding length + total end clearances

Number of gaps = nearest integer of, total gap/porcelain height

Number of packets = number of gaps +1

Height of packet = (leg length -Total gap)/number of packets

All these formulae have been inserted into an EXCEL spread sheet, and typical outputs are shown for various arrangements.

Magnetically balanced cores.

Arrangement for Trefoil type
The core and bottom yoke of a 30 MVAr – 13 kV radially
laminated gapped core reactor. The individual core packets
are bonded with adhesive and bound with glass fibre tape.
Gap spacing is maintained by porcelain spacers.

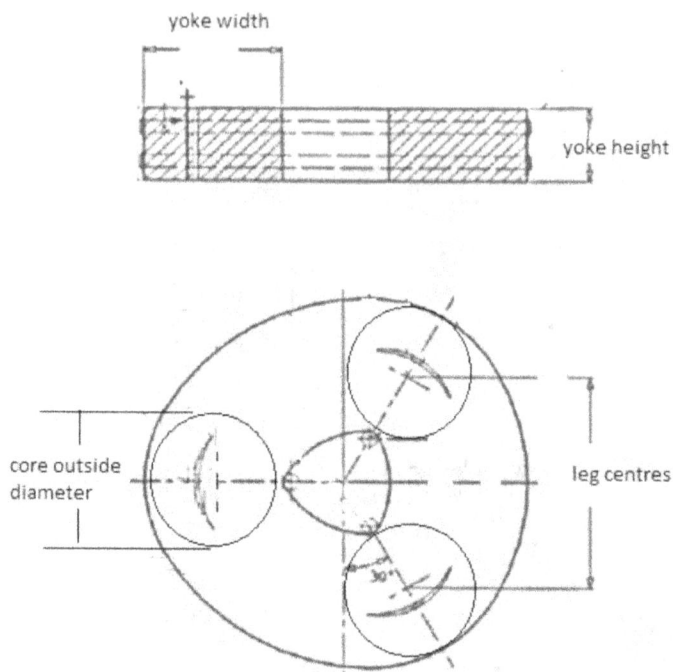

Typical arrangement of Trefoil Yokes

Core and windings of a 60 MVAr 362 kV Trefoil shunt Reactor.

Trefoil Shunt Reactor					
Input		**Data**		**calculations**	
alpha guess	0.370	wdg height(m)	1.680	alpha	0.38
MVAr	60.000	c-w clearance(m)	0.155		
line kV	362.00	topb clear(m)	0.079	Biron(T)	1.838
Biron(T)	1.568	Bot clear(m)	0.079		
core od(m)	0.560	Space Factor	0.320	leglth(m)	1.803
core id(m)	0.160	wdgid (m)	0.870		
flux in iron	0.318	wdg rad (m)	0.142	yoke centres(m)	1.38
core area(sqm)	0.203	wdg od(m)	1.154	kg per yoke	5284.6
freq	50.000	ph-ph (m)	0.228	ohms/ ph	2371.1
Nturns	2227	Aoil(sqm)	0.347	kg/leg	2310.2
amps	95.7	Awdg(sqm)	0.279	Fe weight (kG)	17500
A/mm2	2.79				
disc height(m)	0.035	Astr(sqm)	0.626	stray flux	0.141
gap area(sqm)	0.248			alpha'	0.48
Bgap(T)	1.284	Ht/pkt (m)	0.160	Yoke are sqcm	1393.9
total gap(m)	0.295			yoke width(mm)	559.4
nspaces	8	clamp load/ph (Tonnes)	87.8	yoke height(mm)	259.56
React ohms/ph	2184.1	stamp/circle	24	stamp width (mm)	30
Inductance(H)	6.951	conductor bare axial(mm)	9	bare radial (mm)	2
Bstr(T)	0.226	covered axial (mm)	11	covered rad (mm)	4
volts / turn	93.9	number of conductors	2	conductor area	35
Iron loss (kW)	35.00	amps/sqmm	2.74	Stamp thk	5.5
Tank + clamp	15	MT (m)	3.18	length/leg	7081.9
w/kg Fe	2	Cu weight/trans (kg)	6628	resistance	4.4008
Total load loss	190.9	I^2R (kw)	40.3		
		Stray %	17	Clamp tons/ph	63.89
		Total loss/leg (kw)	47.0		
		number of section axially	101	Core +wdg kG	28953
		cooling area cm²	9E+05	watts/sqcm	0.05
		Gradient	7		

60 MVAr 362 kV

Testing a 75 MVAr 400 kV Trefoil Reactor via a 13/275/400 kV auto-transformer. This unit is 60 Hertz for a site in Mexico.

Trefoil Shunt Reactor					
Input		Data		calculations	
alpha guess	0.250	wdg height(m)	1.857	alpha	0.65
MVAr	75.000	c-w clearance(m)	0.127		
line kV	400.000	topb clear(m)	0.080	Biron(T)	1.522
Biron(T)	1.568	Bot clear(m)	0.079		
core od(m)	0.560	Space Factor	0.258	leglth(m)	1.981
core id(m)	0.160	wdgid (m)	0.814		
flux in iron	0.318	wdg rad (m)	0.185	yoke centres(m)	1.41
core area(sqm)	0.203	wdg od(m)	1.185	kg per yoke	5401.2
freq	60.000	ph-ph (m)	0.228	ohms/ ph	2600.25
Nturns	2248	Aoil(sqm)	0.273	kg/leg	2518.9
amps	108.3	Awdg(sqm)	0.381	Fe weight (kG)	18359.1
A/mm2	2.74				
disc height(m)	0.035	Astr(sqm)	0.654	stray flux	0.152
gap area(sqm)	0.248			alpha'	0.62
Bgap(T)	1.284	Ht/pkt (m)	0.155	Yoke are sqcm	1393.9200
total gap(m)	0.337			yoke width(mm)	559.4
nspaces	10	clamp load/ph (Tonnes)	60.2	yoke height(mm)	259.5638
React ohms/ph	2133.333	stamp/circle	24	stamp width (mm)	30
Inductance(H)	5.658	conductor bare axial(mm)	10.16	bare radial (mm)	2
Bstr(T)	0.233	covered axial (mm)	12.56	covered rad (mm)	4.432
volts / turn	102.8	number of conductors	2	conductor area	39
Iron loss (kW)	36.72	amps/sqmm	2.75	Stamp thk	5.5
Tank + clamp	15	MT (m)	3.14	length/leg	7057.1
w/kg Fe	2	Cu weight/trans (kg)	7456	resistance	3.8847
Total load loss	213.3	I²R (kw)	45.5		
		Stray %	18	Clamp tons/ph	63.89
		Total loss/leg (kw)	53.9		
		number of section axially	102	Core +wdg kG	30977.68
		cooling area cm²	1185088	watts/sqcm	0.05
		Gradient	6		

75 MVAr 400 kV

Trefoil Shunt Reactor					
Input		Data		calculations	
alpha guess	0.150	wdg height(m)	1.060	alpha	0.20
MVAr	15.000	c-w clearance(m)	0.085		
line kV	132.0	topb clear(m)	0.066	Biron(T)	1.417
Biron(T)	1.340	Bot clear(m)	0.066		
core od(m)	0.504	Space Factor	0.400	leglth(m)	1.289
core id(m)	0.144	wdgid	0.674		
flux in iron	0.220	wdg rad	0.091	yoke centres(cm)	94.47
core area(sqm)	0.164	wdg od(m)	0.857	kg per yoke	2925.7
freq	50.000	ph-ph (mm)	88.000	ohms/ ph	1209.3
Nturns	1399	Aoil(sqm)	0.154	kg/leg	1240.6
amps	65.6	Awdg(sqm)	0.132	Fe weight (kG)	9573.1
A/mm2	2.37				
disc height(m)	0.035	Astr(sqm)	0.286	stray flux	0.044
gap area(sqm)	0.202			alpha'	0.21
Bgap(T)	1.086	Ht/pkt (m)	0.183	Yoke are sqcm	1129.1
total gap(m)	0.156			yoke width(mm)	503.4
nspaces	4	clamp load/ph (Tonnes)	42.2	yoke height(mm)	233.64
React ohms/ph	1161.6	stamp/circle	16	stamp width	30
Inductance(H)	3.697	conductor bare axial	7.1	bare radial	2
Bstr(T)	0.154	covered axial	8.1	covered rad	3
volts / turn	54.5	number of conductors	2	conductor area	28
Iron loss (kW)	12.45	amps/sqmm	2.38	Stap thk	4
Tank + clamp	3	MT	2.40	length/leg	3364.6
w/kg Fe	1.3	Total weight/trans	2484	resistance	2.6503
Total load loss	53.3	I^2R	11.4		
		Stray %	11	Clamp tons/ph	37.75
		Total loss/leg	12.6		
		number of section axially	87		
		cooling area cm²	380446	watts/sqcm	0.03
		Gradient	4		

15 MVAr 132 kV

Trefoil Shunt Reactor					
Input		Data		calculations	
alpha guess	0.180	wdg height(m)	1.646	alpha	0.18
MVAr	40.000	c-w clearance(m)	0.080		
line kV	132.00	topb clear(m)	0.060	Biron(T)	1.492
Biron(T)	1.380	Bot clear(m)	0.058		
core od(m)	0.560	Space Factor	0.390	leglth(m)	1.729
core id(m)	0.160	wdgid (m)	0.720		
flux in iron	0.280	wdg rad (m)	0.104	yoke centres(m)	1.01
core area(sqm)	0.203	wdg od(m)	0.928	kg per yoke	3861.8
freq	60.000	ph-ph (m)	0.082	ohms/ ph	454.20
Nturns	893	Aoil(sqm)	0.160	kg/leg	2271.1
amps	175.0	Awdg(sqm)	0.163	Fe weight (kG)	14537
A/mm2	2.34				
disc height(m)	0.035	Astr(sqm)	0.323	stray flux	0.055
gap area(sqm)	0.248			alpha'	0.20
Bgap(T)	1.130	Ht/pkt (m)	0.185	Yoke are sqcm	1393.9
total gap(m)	0.246			yoke width(mm)	559.4
nspaces	7	clamp load/ph (Tonnes)	57.9	yoke height(mm)	259.56
React ohms/ph	435.60	stamp/circle	20	stamp width (mm)	30
Inductance(H)	1.155	conductor bare axial(mm)	7	bare radial (mm)	1.4
Bstr(T)	0.169	covered axial (mm)	8	covered rad (mm)	4.18
volts / turn	85.3	number of conductors	8	conductor area	76
Iron loss (kW)	18.90	amps/sqmm	2.30	Stamp thk	3
Tank + clamp	3	MT (m)	2.59	length/leg	2312.5
w/kg Fe	1.3	Cu weight/trans (kg)	4713	resistance	0.6599
Total load loss	86.4	I²R (kw)	20.2		
		Stray %	6	Clamp tons/ph	49.47
		Total loss/leg (kw)	21.5		
		number of section axially	149	Core +wdg kG	23100
		cooling area cm²	8E+05	watts/sqcm	0.03
		Gradient	3		

40 MVAr 132 kV

Trefoil Shunt Reactor						
Input		**Data**		**calculations**		
alpha guess	0.130	wdg height(m)	1.050	alpha	0.16	
MVAr	25.000	c-w clearance(m)	0.035			
line kV	13.800	topb clear(m)	0.050	Biron(T)	1.474	
Biron(T)	1.440	Bot clear(m)	0.040			
core od(m)	0.504	Space Factor	0.500	leglth(m)	1.105	
core id(m)	0.144	wdgid (m)	0.574			
flux in iron	0.236	wdg rad (m)	0.095	yoke centres(m)	0.80	
core area(sqm)	0.164	wdg od(m)	0.763	kg per yoke	2475.3	
freq	60.000	ph-ph (m)	0.036	ohms/ ph	8.02	
Nturns	115	Aoil(sqm)	0.056	kg/leg	1141.7	
amps	1046.0	Awdg(sqm)	0.123	Fe weight (kG)	8375.6	
A/mm2	2.43					
disc height(m)	0.035	Astr(sqm)	0.179	stray flux	0.037	
gap area(sqm)	0.202			alpha'	0.16	
Bgap(T)	1.167	Ht/pkt (m)	0.147	Yoke are sqcm	1129.1	
total gap(m)	0.184			yoke width(mm)	503.4	
nspaces	5	clamp load/ph (Tonnes)	45.7	yoke height(mm)	233.64	
React ohms/ph	7.618	stamp/circle	16	stamp width (mm)	30	
Inductance(H)	0.020	conductor bare axial(mm)	6.5	bare radial (mm)	2	
Bstr(T)	0.205	covered axial (mm)	14.05	covered rad (mm)	19.65	
volts / turn	69.0	number of conductors	34	conductor area	429	
Iron loss (kW)	15.91	amps/sqmm	2.44	Stamp thk	4	
Tank + clamp	3	MT (m)	2.10	length/leg	242.5	
w/kg Fe	1.9	Cu weight/trans (kg)	2787	resistance	0.0123	
Total load loss	66.4	I²R (kw)	13.4			
		Stray %	18	Clamp tons/ph	43.60	
		Total loss/leg (kw)	15.8			
		number of section axially	57	Core +wdg kG	13395	
		cooling area cm²	2E+05	watts/sqcm	0.07	
		Gradient	9			

25 MVAr 13.8 kV

Trefoil Shunt Reactor

Input		Data		calculations	
alpha guess	0.100	wdg height(m)	1.120	alpha	0.07
MVAr	15.000	c-w clearance(m)	0.035		
line kV	15.000	topb clear(m)	0.055	Biron(T)	1.467
Biron(T)	1.390	Bot clear(m)	0.055		
core od(m)	0.504	Space Factor	0.485	leglth(m)	1.195
core id(m)	0.144	wdgid (m)	0.574		
flux in iron	0.228	wdg rad (m)	0.060	yoke centres(m)	0.73
core area(sqm)	0.164	wdg od(m)	0.694	kg per yoke	2259.3
freq	50.000	ph-ph (m)	0.036	ohms/ ph	15.17
Nturns	160	Aoil(sqm)	0.056	kg/leg	1300.3
amps	577.4	Awdg(sqm)	0.069	Fe weight (kG)	8419.6
A/mm2	2.85				
disc height(m)	0.035	Astr(sqm)	0.125	stray flux	0.018
gap area(sqm)	0.202			alpha'	0.08
Bgap(T)	1.126	Ht/pkt (m)	0.203	Yoke are sqcm	1129.1
total gap(m)	0.146			yoke width(mm)	503.4
nspaces	4	clamp load/ph (Tonnes)	45.3	yoke height(mm)	233.64
React ohms/ph	15.000	stamp/circle	16	stamp width (mm)	30
Inductance(H)	0.048	conductor bare axial(mm)	5.8	bare radial (mm)	2
Bstr(T)	0.147	covered axial (mm)	12.75	covered rad (mm)	11.5
volts / turn	54.0	number of conductors	18	conductor area	203
Iron loss (kW)	10.95	amps/sqmm	2.85	Stamp thk	4
Tank + clamp	3	MT (m)	1.99	length/leg	319.1
w/kg Fe	1.3	Cu weight/trans (kg)	1732	resistance	0.0342
Total load loss	50.4	I^2R (kw)	11.4		
		Stray %	7	Clamp tons/ph	40.62
		Total loss/leg (kw)	12.2		
		number of section axially	66	Core +wdg kG	12182
		cooling area cm²	2E+05	watts/sqcm	0.08
		Gradient	10		

15 MVAr 15 kV

The complete core and winding assembly of a 60 MVAr radially laminated gapped core shunt reactor. The yokes are of silicon iron strip. This unit is provided with on load tap changing equipment for a rating variation of from 30 to 60 MVAr in sixteen steps.

The complete shield of a single phase 33.3 MVAr 380 kV magnetically shielded shunt reactor. The ducts in the shield are provided to increase the shield depth to that of the diameter of the winding. They also facilitate the clamping of the winding.

The completed shield and winding of the unit shown. The clamping of the shield and winding is provided by circular steel plates and glass fibre tie rods which pass through the yoke ducts. The major insulation between the winding and the shield consist of a series of barriers which are clearly visible.

In some cases, where there are no suitable high voltage test transformers available, the reactor may be fitted with temporary test legs, wound with low voltage windings as shown. These are then fitted inside the electrostatic screen, and the unit can be tested by inducing the test voltage in the main winding by normal transformer action.

References

There have been many publications on transformers and reactors, and over the years, I have attended many colloquium and conferences, including CIGRE, IEEE, CEPSI and EUROCON, and from theses I have selected many published papers that are mainly associated with transformers and reactors. During my time as technical manager with Peebles, and VA Tech, I have produced a number of training discs that have been distributed throughout the group. These training discs; called TALK discs, contained these selected papers and also contained some Power Point presentations, and Transformer Topics, which were distributed by Peebles over many years.

Each TALK disc has a list of the contents in WORD form, and as there are twenty discs, it would be too extensive to include in this book.

There is a CIGRE document; 'GUIDELINES FOR CONDUCTING DESIGN REVIEWS FOR TRANSFORMERS 100 MVA AND 123 KV AND ABOVE', The principles of this may be used to conduct any design review. However, for shunt reactors there are fundamental differences in the operation and these must be taken into account.

In normal circumstances the design review should be held in the manufacturer's workplace, with input from the department of engineering.

The seller should lead the review with a presentation of the product. It should be an open book session with any questions from the buyer being answered fully. The seller should minute the meeting and action any points that require clarification.

The buyer should be aware of the salient design features, and ask questions on what procedures are in place to mitigate for any concerns regarding new developments. This should include detailed questions on manufacturing and processing procedures including all the critical inspection stages.

The buyer should also discuss any quality plans used by the seller, including inspection of any bought out items. A full test procedure should be made available, an assurance should be given that the appropriate test plant is in place to perform these tests fully.

There will soon be another Brochure issued by CIGRE on REACTORS, and this document will be very detailed.

Acknowledgements

I would like to thank Professor Sinclair Gair for reading this book and his valuable contribution and suggestions for improvement. It has involved several revisions; each one requiring a re-read to keep improving the layout.

I would also like to thank my wife Joyce for her patience and tolerance, giving me the time and space to write.

www.ingramcontent.com/pod-product-compliance
Lightning Source LLC
Chambersburg PA
CBHW072306210326
41519CB00057B/2957